Happy Bir
D0149436
April 4, 2006
Love, Dad & Toni

The MOMMY BRAIN

ALSO BY KATHERINE ELLISON

The New Economy of Nature

Imelda

The MOMMY BRAIN

How Motherhood Makes Us Smarter

Katherine Ellison

A Member of the Perseus Books Group
New York

Published by Basic Books,
A Member of the Perseus Books Group

Books published by Basic Books are available at special discounts for bulk purchases in the United States by corporations, institutions, and other organizations. For more information, please contact the Special Markets Department at the Perseus Books Group, 11 Cambridge Center, Cambridge MA 02142, or call (617) 252-5298 or (800) 255-1514, or e-mail special.markets@perseusbooks.com.

Designed by Brent Wilcox
Text set in 11.25 point New Caledonia

Library of Congress Cataloging-in-Publication Data
Ellison, Katherine, 1957–
The mommy brain : how motherhood makes us smarter / Katherine Ellison.
p. cm.
Includes bibliographical references and index.
ISBN 0–465–01905–6
1. Mothers—Psychology. 2. Mothers—Intelligence levels. 3. Motherhood. 4. Intellect. I. Title.
HQ759.E463 2005
306.874'3—dc22

2004026506

05 06 07 / 10 9 8 7 6 5 4 3 2 1

For Bernice

CONTENTS

Part One
THE GREAT TRANSITION

CHAPTER

1

Smarter Than We Think

Smart\smart\adj 1: making one smart: causing a sharp stinging 2: marked by often sharp forceful activity or vigorous strength (a ~ pull of the starter cord) 3: BRISK, SPIRITED 4 a: mentally alert: BRIGHT.

MERRIAM WEBSTER'S
COLLEGIATE DICTIONARY

A FEW WEEKS AFTER my first son was born, I had a troubling dream. It was September 1995, and I was on leave from my job as a foreign correspondent in Rio de Janeiro. In my nightmare, space aliens had landed in Brazil's capital, Brasilia, but I stayed home, unable to decide whether the story was worth pursuing. The dream later struck me as the perfect showcase for my fear that I'd traded in my brain for my new baby.

It was just that fear that had kept me and so many of my peers from having babies at all, right up until we'd almost lost the chance to choose. The problem was that I'd come to depend upon my brain for so many good things, including my livelihood, my self-esteem, and my freedom to marry for love. And I knew that becoming a mother made me subject to a modern affliction called *Mommy Brain*—which, like "senior moment" is a cheery synonym for abrupt mental decline. The phrase summons the image of a ditsy pregnant woman who weeps at tissue commercials, or of a frazzled mom with nothing in her head but carpool schedules and grocery lists. ("If you've left the crayons to melt in the car / And forgotten just where the car keys are / There's a perfectly good way to explain: / You see, you've come down with "Mommy Brain," reads a poem by one self-alleged victim.)

Along with varicose veins and thickened waistlines, diminished cerebral capacity would appear to be a risk inherent in women's reproductive fate. That's certainly how many nonparents perceive pregnant women and new mothers. When researchers showed audiences videotapes of a woman in various workplace situations—the same woman, the same work, but in some scenes wearing a prosthesis so that she'd appear pregnant—the "pregnant" woman was rated less competent and less qualified for promotion. We mothers also perpetuate this bias. "Mommy Brain!" is our frequent alibi when we say something dumb. "Part of your brain exits with the placenta!" one friend advised me early on.

The pessimistic chorus wasn't always this loud. The phrase "Mommy Brain," which is of relatively recent vintage, followed the historic flood of women into the workplace beginning in the 1960s. This change brought new scrutiny from others—and a new self-consciousness for mothers. Today nearly three-fourths of mothers with children aged one or older are at work outside the home, frequently in jobs requiring mental sharpness, making many of us more vigilant than ever before about fluctuations in our mental acuity. And not only do our jobs require more brain power; rearing children today amidst information overload and furious debates over nearly every aspect of parenting takes more smarts than ever.

Now, few moms would deny that children challenge our mental resources. The hormonal roller-coaster, sleep deprivation, biased bosses, brainless chores, and too much Raffi are just part of the toll. Because men, despite some notable recent progress, still aren't equitably sharing these burdens, we're left with a mostly female predicament. But what makes it all harder is a residue of feminism. The same fierce rhetoric that gave women the courage to brave an unwelcoming job market created a harrowing "Mommy Brain" image for today's mothers, myself included, who were then coming of age.

In 1963, in *The Feminine Mystique,* Betty Friedan compared women who devote themselves to the home to "walking corpses." Such women, she wrote, "have become dependent, passive, childlike; they have given up their adult frame of reference to live at the lower human level of food and things. The work they do does not require adult capabilities; it is endless, monotonous, unrewarding." A few years later, movie goers and novel readers would meet the vivid embodiment of Friedan's brain-dead momma in Tina, the dithering, pill-popping heroine of a best seller aptly titled *Diary of a Mad Housewife*.

The doom-saying didn't end with the last century. It remains a private and surprisingly frequent public refrain today. "Anyone who tells you that having a child doesn't completely and irrevocably ruin your life is lying," muses the character Julie Applebaum, who, in *Nursery Crimes*, the 2001 novel written by the retired public defender Ayelet Waldman, gives up a career as a public defender to stay home with her new daughter. "Everything changes. Your relationship is destroyed. Your looks are shot. Your productivity is devastated. And you get stupid. Dense. Thick. Pregnancy and lactation make you dumb. That's a proven scientific fact."

It's far from "scientific fact," as we shall see. But this sort of stuff is discouraging to read if you happen to be a mother. So is the following self-deprecating comment made by *Newsweek* columnist Anna Quindlen as she reflects in 2004 on her own reproductive transition: "It was as though my ovaries had taken possession of my brain. Less than a year later an infant had taken possession of everything else. My brain no longer worked terribly well, especially when I added to that baby another less than two years later, and a third fairly soon after that."

During those same years, it's worth noting, Quindlen won a Pulitzer Prize for commentary in the *New York Times* and wrote several successful novels and advice books. No small accomplishments for this mother of three. Yet for some reason, Quindlen feels obliged to assure readers that motherhood has dulled her intellect.

Maybe she's just bowing to peer pressure. Polls in recent decades have tracked a marked decline in many parents' satisfaction with the job of rearing children, a trend owing greatly to the perceived price we pay. Complaining about what our children have done to our finances, moods, hips, and *brains* has become a fashionable pastime at parties as well as

, theme of several recent books. *Senility is something you inherit from* *ur kids,* we joke. But the new parental angst is serious, and no doubt ɔart of the reason so many women have delayed childbearing right up to the brink of menopause.

I got in just under the wire. By the time I gave birth, at what my obstetrician politely called my "advanced maternal age," I'd waited so long that it was already hard to say whether "Mommy Brain" or early onset of senility was more to blame for my occasional mental lapses. Joey was born when I was thirty-eight years old, Joshua three years later. I knew I was taking the risk of never having children by waiting so long. But I feared that brain damage might cost me the job I'd wanted ever since I was a child.

I was raised in the suburbs, the youngest of four children; my parents were a physician and his stay-at-home wife, a college beauty queen who had dropped out of school to marry. We called my mother "the geisha" when we weren't calling her "the martyr." The family legend was that her fate, and ours, depended upon my father's brilliance. Yet, as I realized only much later, the very perpetration of this legend proved my mother's smarts. She worked under the radar to accomplish her goals, networking at a furious pace to establish her family in the community and further her children's prospects. She waited until I had left for college before earning her own degree; and for ten years thereafter, she taught elementary school children afflicted with learning disabilities.

Although my mother's personal example implied that women's chief priority is to serve their families, she not only took pride in her two daughters' achievements but also encouraged our career plans. We took this for granted, assuming that, unlike her, we were too smart to waste our time cooking and cleaning. All my siblings became medical doctors, but I left the fold early on. At sixteen, I traveled to Nicaragua, then ruled by Anastasio Somoza, as an Amigos de las Americas medical volunteer. I was shocked to learn of my government's support for a dictator who was stealing humanitarian aid and stifling dissent. If more Americans knew, I thought, the support would have to end.

I returned home determined to become a foreign correspondent, and five years later I was hired at the *San Jose Mercury News.* Soon, I was reporting from Central America, a job that produced one major collateral benefit: In 1982, in a government press room in Managua, I met the man I would eventually marry. Jack was a freelance writer traveling through Nicaragua, and we courted for the next eight years before marrying and settling in Rio, when the *Miami Herald* hired me as their correspondent. Three years later, I was pregnant with Joey.

As I watched my body morph, I prepared for more permanent changes. For most of my life, I'd enjoyed the control and freedom that come with an observer's point of view. Motherhood, I suspected, would cost me a lot. And I was right. But it was then still impossible to imagine what I would gain.

We stayed in Rio for the next four years. In 1999, we moved back to the San Francisco Bay area, one year after the birth of Joey's brother, Joshua. Jack quit freelancing in return for a steady job, and I quit the *Herald* to write a book about environmental conservation. In the process, we switched from our previous Brazilian model of nanny-supported childcare to the contemporary U.S. suburban style, which meant that I would try to do everything at once.

This at last was true Mommy Brain terrain, a land of 24/7 distractions, silly music, and such bleakly repetitive duties as wiping pee from the toilet seat. My psychiatrist sister, Jean, whose children by that time were in college, understood my distress when she called one night as I was simultaneously trying to cook dinner and break up a fight over a Pokemon card, an AT&T computer technician on call-waiting. "Don't worry," she said, responding to the shrill pitch of my greeting. "The damage isn't permanent."

But by then, I'd already come to a startling conclusion. I didn't feel particularly damaged, after all. True, I was complaining a lot more. But I was also accomplishing more. Though I often felt frazzled, I was more motivated, excited by all I was learning at work and at home. My children not only had inspired my future-oriented interest in the environment but also had provided me with the "excuse" to insist on a more flexible work life; this, in turn, allowed me more creativity. The children were also giving me constant lessons in human nature: theirs and my own.

Although I'd had newspaper deadlines before, never had I faced the unparalleled urgency of a baby who needed to breastfeed, or a preschool teacher at close of day, both of which taught me a new kind of focus. Within two years of our move to California, despite constant interruptions, I had finished my book, gone on a speaking tour, launched a freelance career, helped my kids adjust to a new community, supervised repairs to our home, found a great circle of friends, and tracked down a qualified expert to help a babysitter afflicted with early-stage leprosy. I had many more reasons for worry, yet, to my surprise, I felt calmer. And I kept running into other mothers who felt the same way.

Could I have entered this phase of more professional fulfillment and lasting relationships if I had not had children? Was it all just a function of the purported wisdom that comes with age? I don't think so. Instead, I was beginning to believe there was more to the Mommy Brain than I had ever imagined. *Maybe it wasn't all bad news.* And so, in time filched from freelancing, housework, and childcare, I began to probe beyond the cliché.

I was encouraged early on by a report I'd read in 1999 about two Virginia neuroscientists and researchers, Craig Kinsley and Kelly Lambert. They had compared the performance of mother and bachelorette rats on a learning and memory test and discovered that the mothers led the pack. Further, these learning and memory advantages appeared to last long into the rats' golden years, well after they'd stopped reproducing. When the two researchers published their results in the prestigious journal *Nature* that year, they generated a small burst of publicity, including one headline that boldly declared: "Motherhood Makes Women Smarter."

As I continued to dig, I found that Kinsley and Lambert weren't alone in their perception of a transformed, and even improved, Maternal Brain. Eventually, I interviewed dozens of scientists in the United States and overseas, many of whom have been generating evidence powerful enough to eradicate the Mommy Brain stigma. My excitement grew as I realized they were forging a new frontier of knowledge that was just as dynamic and far-reaching as the previous decade's study of gender differences.

In contrast to the traditional "parenting studies" focused on the child, this newer work investigates the impact of parenthood on parents. It revolves around one pivotal idea: By means of a dynamic combination of love, genes, hormones, and practice, the female brain undergoes concrete and likely long-lasting changes through the process of giving birth and raising children. It's a transformation on as grand a scale as puberty and menopause, though in the past it has rarely been treated as such. Yet what is especially moving about the transition to motherhood is that it does not take place in isolation, as do puberty and menopause, but happens as part of an enduring relationship—the most passionate crucible of a relationship there is. For the two minds immediately involved, and all the other lives those minds eventually touch, the consequences can be profound.

We are all used to hearing that motherhood is a time when "everything changes." Yet the idea that motherhood concretely changes our brains seemed strange and wonderful to me. In February 2003, I published a feature story based on my preliminary research in *Working Mother* magazine. I described how the Mommy Brain phenomenon may be part of our modern social experiment in which women, more than ever, are trying to have their minds in two places at once. I also recounted some of the condition's well-known challenges and lesser-known strengths. The reaction convinced me I'd hit a nerve. "Thank you!" one working mother wrote. "Your article has put my concerns at bay . . . while I may sometimes mail my credit card payments to the wrong address, I can in one instant quote test results to a client, while in another remember where my daughter left her purple Barbie shoe." I was also finding that every time I mentioned the phrase, "what motherhood does to your brain," the topic ignited huge interest and debate. Suspecting that I had scratched the surface of a new and largely untold scientific story, I decided to write this book.

As I interviewed scientists in the months that followed, I combined my research with my own hands-on experiments in rearing my two young sons, Joey and Joshua. My boys have variously been described as "high-spirited," "high-energy," and "high-maintenance" by their teachers, and as *vilde chayas,* Yiddish for wild animals, by my parents. Yet, in ways I felt driven to explore, my children kept leaving me not just physically spent but mentally energized. At the same time, leading scientists I spoke with were confirming my suspicion that motherhood might actually help *improve* the mind. "From a neurological viewpoint, it's a revolution for the brain when you have a child," says Michael Merzenich, a pioneering expert on brain development at the University of California at San Francisco. "It is life-changing in the sense that you are presented with physical, mental, mechanical challenges—forty-nine disasters to take care of at once. It's an epoch of learning and brain-induced changes, because everything matters so much. . . . I don't think there are a lot of better things you can do for your brain than have a child."

Motherhood is far too complex and variable a condition for anyone to argue that mothers, as a rule, are smarter than women who have not given birth. Furthermore, most advantages gained from the experience depend not only on your circumstances but on your attitude. If you're under debilitating stress, for instance, you're obviously likely to miss out on many benefits. Still, after what I've learned, I feel confident in proposing that a Mommy Brain should be thought of less as a cerebral handicap and more as an advantage in the lifelong task of becoming smart.

By "smart" I mean much more than the ability to multiply two-digit numbers in your head. I have in mind instead the kind of "bright" that translates into enhanced perception, efficiency, resiliency, motivation, and social skills (or "emotional intelligence"). In these first-order survival capacities—which I'm calling the five attributes of the baby-boosted brain—mothers' capacities can be most strengthened, a process I detail in the heart of this book.

In the pages that follow, I also take you with me on my quest to track down what we know about the powerful mental advantages that motherhood can bestow. You'll witness in Virginia revolutionary lab experiments on brainy mother rats; at Yale University, sophisticated functional magnetic resonance imaging (*f*MRI) exams tracking the brain behavior of new parents; and in California's Silicon Valley, an innovative childcare project that helps mothers be their smartest at work.

Throughout this book, my emphasis will be on mothers, because mothers clearly undergo the most dramatic physical changes in giving birth, and because, throughout human history, mothers have devoted much more of their time and attention to their children than anyone else. (In 2004, the first U.S. government survey to quantify this well-known fact found that the average contemporary working woman was spending about twice as much time as the average working man on household chores and childcare.)

Furthermore, recent research, described in Chapter 3, suggests that mothers may be—as has long been suspected—more "hard-wired" to respond to their infants. Even so, Mommy Brain gains aren't limited to mommies, and I show how fathers, other caregivers, and every day altruists can also partake of the advantages of caring proximity to children. I also reveal how some particularly smart mothers and managers have made the most of parents' special skills in the workplace. Along the way, I offer insights from mothers I've interviewed—including mothers who are also professional scientists, and who have turned their informed and painstaking attention onto their own experiences. Several of these experts share their advice about how to make the most of your own Mommy Brain.

That most women have the potential to become smarter with motherhood may turn out to be one of this century's most hopeful ideas. It's time for mothers to recognize that—as Joshua once whispered in my ear, to my amazed pride (until I learned he'd stolen it from *Pooh's Grand Adventure*)—"You're stronger than you seem, and smarter than you think."

CHAPTER
2

"Honey, the Kids Shrunk My Brain!"

Whatever does not kill me makes me stronger.
FRIEDRICH WILHELM NIETZSCHE

ALTHOUGH IT'S NEARLY impossible to convey the animal shock of childbirth to someone who hasn't experienced it, the writer Rahna Reiko Rizzuto comes close:

> After an hour of watching my hands twist while searching for my own painkilling sound, my water broke and I discovered just how slowly I could form the thought, What in the hell was that? Let me illustrate: First, something shuddered inside me and I heard the far-off sound of, say, a potato exploding in a microwave oven, and I thought, What in—. Then I felt my underwear bulge and I got as far as the hell—. Then the baby's no-longer-pillowed head came crashing down on the bundle of nerve endings in my tailbone and I thought, was that? just as the amniotic fluid turned into a small lake around Craig's feet and he said, "What in the hell was that?"

Some variation of this combination of combustion and loss of control is how most of us join the club of motherhood, with its seemingly rapid progress from fretting over what the hell happened to our mucous plugs to what the hell our son is doing, still out with the car at 3:00 A.M. And there's more to this tale of woe. Combine the chemical chaos of pregnancy and labor with sleep deprivation and the frantic task of learning a whole new curriculum in an extremely short time—with absolutely *terrible* consequences if you relax your guard, even for an instant—and it's easy enough to understand why pregnant women and new mothers occasionally plead not guilty to any number of mental failings by reason of Mommy Brain. The condition is known as "maternal amnesia," and even "pregnancy dementia" in some scientific circles, "porridge brain" in the United Kingdom, "placenta brain" in Australia, and "Mammy Brain," as one scientist translated it, in Japan.

The idea that parenthood depletes us is hardly new. "Profusely blinde, we kill ourselves to propagate our kinde," the poet John Donne wrote in 1611. But in recent years, the Mommy Brain notion that babies steal our brain cells has become so compelling that it has led dozens of psychiatrists, psychologists, and endocrinologists around the world to investigate whether it is really true. In 2001, two British neuroscientists, Matthew Brett and Sallie Baxendale, even staked a claim to define a new psychiatric disorder, which they dubbed GMI: Gestational Memory Impairment.

Most pregnant and early postpartum women who have been surveyed to date certainly *believe* they've been impaired, a belief that often brings with it a great deal of shame and fear. Self-alleged victims report memory lapses, distractibility, weak concentration, and "general cognitive slowing." The neuroscientist Jeffrey Lorberbaum, who has brain-scanned new mothers as part of his research at the Medical University of South Carolina, is one of the many researchers who have heard these complaints first-hand from volunteer subjects. "Universally, without exception," he says, "they say their brains have turned to Jell-O."

Despite this popular conception, however, scientists who have tried to establish proof of an actual Mommy Brain mental handicap have come up with strikingly varied and ultimately inconclusive results. Although two small but well-publicized studies of pregnant and early postpartum women reported evidence of minor memory problems, several others found no change, and one showed an improvement.

This discrepancy could have any one or more of three explanations. Some powerful aspects of giving birth and coping with a newborn do mentally stun us, at least temporarily. However, some scientists speculate that pregnant women and new mothers who have indeed become more distracted can focus enough to ace a cognitive laboratory exam—thus explaining why some tests reveal no problems. Nevertheless, as I describe in this chapter, uniquely contemporary challenges and certain cultural perceptions that put mothers at unnecessary disadvantage may well be the overriding reasons why pregnant women feel less capable than usual, even when so little evidence exists to show that they really are.

In the end, I'm certain that the belief in a Mommy Brain as an *impaired* brain is dangerously misleading. What's really going on, some experts suggest, is much more complex—and encouraging. A pregnant and early postpartum woman's brain is tied up in a major, hormone-powered transition, a process that Craig Kinsley, a neuroscientist the University of Richmond, Virginia, calls a "reorganization." Any experience of temporary ditziness, he says, "is a tradeoff for a better functioning and focused brain later on."

Motherhood—just like puberty—may thus knock us off our feet for a time, only to set us back up, often stronger than before. "It all relates to the brain maximizing attachment to the young without much room for error," theorizes the neuroscientist Michael Merzenich. "It's analogous to the situation that arises whenever you're in a very high-stakes, life-threatening situation. The brain is specialized for quick, decisive action, at the expense of cogitation or learning. In a sense, the brain doesn't have *time* for complex cognitive stuff. It's all about protection, nurture, attachment, concentration on the over-riding task—and *focus*!"

Back in 1956, the psychoanalyst and pediatrician Donald Winnicott came up with his own, gentler, name for the yet un-named Mommy Brain phenomenon. He called the intense first weeks of motherhood a time of "primary maternal preoccupation" and went so far as to compare it to the acute onset of mental illness. Winnicott described this condition as one of heightened sensitivity in which the mother focuses on the infant to the apparent conscious exclusion of everything else.

This approximation of insanity is actually essential for mothers to learn all they need to know about the newcomer in their lives, Winnicott theorized, including the baby's needs, his unique temperament, and his unfamiliar way of expressing himself. By the time this has subsided, mothers seem to recall few details of the degree and depth of their preoccupation.

In this context, it's easy to see why the prominent Australian neuroscientist Allan Snyder compares pregnant women with Albert Einstein. "Women's memory is not reduced during pregnancy," he maintains. "Rather their attention is on things that are more immediately crucial. Einstein was known to forget where he put checks of large amounts, not because of a bad memory, but rather because of deep concentration on things of greater importance." Encouraging as this paradigm may be, however, it's important to remember that new mothers are coping with some serious physical challenges that Albert Einstein could barely have imagined.

Shrunken Brains and Heavy Drugs

New mothers and their loved ones may not be surprised to learn that pregnancy *does* shrink your brain—for a few months. In 1997, Anita Holdcroft, an anesthesiologist, and her colleagues at the Royal Postgraduate Medical School in London used magnetic resonance imaging (MRI) technology to scan and measure the brain volume of eight healthy women. Scientists had previously found evidence of reduced brain size in pregnant women suffering pre-eclampsia, a dangerous condition found in more than one in twenty pregnancies, and involving high blood pressure. Holdcroft wanted to see whether that applied to healthy women as well. She found a significant decrease in brain size—almost 7 percent in one volunteer, that reached its peak at the time of the baby's birth and returned to normal within six months later.

One irreverent British headline described the news as proof that: "Baby . . . Is Eating My Brain Cells," and the study's authors offered more sober wording for what perhaps amounted to the same thing. They suggested that some of a pregnant woman's physical resources were being temporarily diverted from the brain, a major energy guzzler, to nurture the growing fetus.

We don't yet have the technology to know precisely what is happening to a human female's brain at this critical time, but we do have a good idea, thanks to rats. When Craig Kinsley and his colleague Kelly Lambert, at Randolph Macon College in Virginia, dissected the brains of late-pregnant rats, they observed a complex remapping of neural pathways in the hippocampus, the center of learning and memory. Neurogenesis—the brain's continuous manufacturing of new cells, or neurons—had slowed down, perhaps helping to account for the brain shrinkage Holdcroft detected. And yet, nerve cells in the hippocampus had sprouted many more dendritic spines.

It's time to interrupt this narrative for some picturesque brain basics. Each *neuron,* or brain cell, has a long trunk and many branches that make it look roughly like a tree at winter's end. The branches are called *dendrites,* which can sprout buds, called the *dendritic spines.* In the center of the branches is the *cell body*; this contains the nucleus and all the other parts needed to keep the neuron alive. The long trunk is the *axon,* a kind of information highway.

Now imagine a wild forest inside your brain, composed of as many as 100 billion of these neurons and their sinuous dendrites. The dendritic spines are almost, but not quite, touching the axons of other neurons. Information—thoughts and feelings—travels along the axons in the form of chemical *neurotransmitters,* which build up until they generate a tiny electrical impulse that carries them across the little gaps to connect with other cells' dendritic spines. The little gaps are called *synapses.*

Each time you think or behave in a new way—for example, each time you worry about your child's welfare, or tell him to look both ways before crossing the street—some of these new connections in your brain are made stronger. These changes occur each time you repeat that thought or action. This is the essence of learning, and the reason for the saying among scientists that "neurons that *fire* together *wire* together."

The meaning of the blossoming of dendritic spines, revealing the creation of many new synapses, that Kinsley and Lambert observed in the pregnant rats' hippocampi remains a matter of speculation. It's possible that such an exuberant growth contributes to the feeling of distractedness that many women report. But Kinsley, sanguinely, compares it to the chaos of a toy factory just before Christmas, or to a computer acquiring extra bandwidth to help it run more than one program at a time. In all these examples, there might be glitches along the way, but great rewards in the long run. Regarding a mother rat and her litter, he and Lambert have written that "neural activity brought about by pregnancy and the presence of pups may literally reshape the brain, fashioning a more complex organ that can accommodate an increasingly demanding environment."

Underlying this transformation is the potent marinade of reproductive hormones bathing the pregnant female's brain. It has been estimated that by the last weeks of pregnancy, levels of three kinds of estrogen rise by up to several hundred times their norm. Progesterone shoots up as much as tenfold, and levels of the stress hormone cortisol can double.

Many scientists assume that something in that marinade fogs a woman's thinking—if only to help her forget the pains of pregnancy sufficiently so as to reproduce again one day. There is disagreement on which ingredient might be most to blame, however, and, despite some circumstantial evidence, no clear understanding of cause and effect.

Liisa Galea, a psychology professor at Canada's University of British Columbia, suspects estrogen may be the major culprit. Galea, who had trouble finding her car in the campus parking lot during her own last weeks of pregnancy, has tested pregnant rats' performance in a water maze. The rats' task was to remember the changing location of a floating platform and crawl to safety. Rats have a three-week-long pregnancy. In the third trimester, when estrogen is highest, the rats clocked in their worst times.

What's interesting here is that a large body of literature suggests that, under the right circumstances, estrogen is a brain *tonic.* Research shows that young adult women feel smarter at the time in their menstrual cycles when their estrogen peaks, and do better at some tasks, such as those requiring verbal fluency. And several trials have demonstrated that estrogen-replacement therapy helps minimize declining verbal memory in postmenopausal women. The hormone is known to play a role in the formation of new synapses, like the ones Kinsley and Lambert observed in the pregnant rats' brains, and in neurogenesis. But because researchers still don't understand the impact of high estrogen levels on memory, Galea speculates that "all those new synapses may temporarily just be leading to more *noise.*"

The jury is clearly still out on estrogen, while some researchers theorize that another hormone, progesterone, may be causing more mental trouble. Those who endorse this view point to a study showing that female volunteers who took oral progesterone to give them blood levels comparable to those in late pregnancy declined significantly in their ability to remember the details of paragraphs read to them. Still another camp of experts suspects that the high levels in pregnancy of the stress hormone cortisol, a glucocorticoid, may cause confusion. Cortisol can make you more alert—it's the hormone involved in the "fight or flight" response. But, as Merzenich suggested, it's also a hormone that focuses the mind on the most important task at hand.

One recent preliminary finding suggests that yet another key factor may have been ignored for years in the search for a smoking gun in the Mommy Brain drama. In late 2004, two researchers from the Simon Fraser University in Canada said they'd found in sophisticated tests that only women carrying *girls* demonstrated cognitive decline. Women carrying boys had no problems. If this finding, unpublished at this writing, is replicated, it could shed new light on the fascinating biological interplay between a mother and her unborn baby.

Shock

Thanks to researchers at Jerusalem's Hebrew University, we have compelling proof that women are not at their intellectual best on their first day postpartum. In 1993, the Israeli scientists published the results of tests they had performed on one hundred new mothers; the women were quizzed through standardized neuropsychological tests, their answers compared with those of nonpregnant childless women, third-trimester high-risk pregnant women, and fathers of newborns. On that first eventful day, the new moms scored significantly lower than the other groups. Yet none of these deficits in verbal and visual recall were apparent on the second or third day.

To this, I can say only "Duh." You've just been through probably the scariest and most exhilarating, painful, and tiring hours of your life, and you are now looking at your future in its yowling red face. It's possible that your performance on standardized neuropsychological tests simply isn't a top priority. I recently looked back at the journal I started the week Joey was born, after thirteen hours of labor. My hope had been to recall all the exquisite details of some of the most strangely euphoric days of my life, and I did find a few. But what popped out at me was a scrawled list of times, down to the minute, that Joey nursed. It was almost every half hour. I had to write it down just to begin to fathom how, for me and this noisy newcomer, previously taken-for-granted tasks of survival had suddenly assumed center stage.

Many mothers understandably feel emotionally and physically vulnerable at this time. As Cort Pedersen, who specializes in postpartum depression at the University of North Carolina at Chapel Hill, says: "That first baby is a plunge into something extraordinary, in that your ability to control life as you had been used to is suddenly gone. You can't pee, can't sleep, can't do anything when you want to, and that is a real shockeroo. Often women aren't prepared. Nobody has talked to them on this level."

There is an emerging literature that tries to do just that: books with titles such as *I Wish Someone Had Told Me.* Yet, no matter how many Lamaze classes you've attended, no matter how many movies you've seen, or dolls you've swaddled, in the end, as a first-time mom, you're as likely as Rahna Rizzuto was to be blindsided by childbirth. For my part, I had been tear-gassed, mugged at knife-point, taken hostage by hysterical Mexican *campesinos*, rear-ended on a freeway, and chased by killer bees. Yet never—until about an hour before my son Joey's head appeared—had I thought my body might actually *implode.*

So perhaps it's no wonder that some form of the "baby blues" affects up to 80 percent of new mothers in the first few hours or days postpartum, according to the American College of Obstetricians and Gynecologists. About 10 percent of mothers suffer a more extreme postpartum depression, one that lasts longer and is more intense.

Some scientists in recent years have even questioned whether childbirth, particularly an extraordinarily difficult delivery, can lead to a variant of Post Traumatic Stress Disorder (PTSD). (The National Institute of Mental Health defines PTSD as arising "after exposure to a terrifying event in which grave physical harm occurred or was threatened." Let's assume the threat of implosion qualifies.) Depression and troubles with concentration or memory are listed as common results of this disorder. Yet both symptoms are usually avoidable or can be minimized with skillful care.

Sleep No More

Of all the potentially mind-wrenching baggage of early parenthood, one of the heaviest loads is doubtlessly sleep deprivation. To keep someone from sleeping is "to tamper with their equilibrium and sanity," according to the psychotherapist John Schlapobersky, who was tortured by the South African government during the apartheid of the 1960s. Sleep deprivation is a well-known torture technique that has been used by military interrogators worldwide. Yet, despite our understanding of how sleeplessness can affect the brain, many new mothers are unprepared to cope with a new baby's impact—even though this toll, too, can be greatly reduced with sufficient preparation and skill.

James Maas, a psychology professor at Cornell University, estimates that in the first year of an infant's life, the primary caregiver loses up to seven hundred hours of sleep. Compounding this problem, Maas says, is that new parents may underestimate the impact of this loss, chalking up sudden mood swings, for instance, to the way a spouse sits reading the newspaper while you're walking the floor with a hysterical baby. Indeed, new moms may wrongly assume that not only their marriages but their brains have been damaged, when all they really need to do is find a way to take more naps. Underestimating the magnitude of the transition to parenthood, many new couples also mistakenly return to work too early or turn down offers of help from friends and family. (More realistic, paid family leaves would also help, a topic I'll discuss in Chapter 12.)

If you persist in your slumber deficit, you may well find yourself morphing from Sleepy to Dopey. The reason is that the brain's frontal cortex, responsible for keeping you alert, undistracted, innovative, and flexible, is the first to falter during extended sleep loss. Laboratory research shows that sleep-deprived volunteers tend to use a reduced vocabulary, with more clichés, and have more trouble finding creative ways to solve complex problems.

Robert Sapolsky, a biology professor at Stanford University and a leading national expert on stress, says it never occurred to him to study sleep deprivation until he had fathered children, after which it seemed central. "Newborn sleep deprivation is the worst kind you can possibly get," he argues. "If you have your total amount of sleep decreased, that's a stressor on the system and impacts your mood and gets you depressed and interferes with cognition. If you have a decreased total amount of sleep that is also fragmented, that's even worse. But the absolute worst is a decreased total amount of sleep which is fragmented *unpredictably*. It's not for nothing that medical residents are psychotic."

The mechanics, says Sapolsky, have to do with those stress hormones—glucocorticoids—and their jarring impact on the brain. Even as we sleep, these hormones are responding to an internal clock. "If you go to sleep where you're expecting to be woken up at 5:00 A.M. you begin to have a rise in stress hormone levels at 4 o'clock, since normally you have them about an hour before spontaneous waking," Sapolsky says. "But if you go to sleep anticipating you could be woken up any second during the night, you're always physiologically preparing for the stressor of waking up." In other words, you could have a normal night's sleep as far as hours are concerned, but you can also be so stressed out that the sleep does you no good.

From my own experience, compounding the impact of knowing you could be woken up at any time is not knowing just how that's going to happen. Someone could be trying to nurse your nose; someone could be poking a finger in your eye; someone could be head-butting you, or, worse, calling you from the side of a highway. My brother Jim told me how his three-year-old son once woke him by lifting up his head in his hands and shouting, "Oh, no!" and scurrying over to his mother.

Although some interference with sleep can't be avoided, you can take measures to reduce the toll. Napping helps a lot, and this is where negotiating with your husband, partner, mother, sitter, neighbor—or even your boss—is essential. (Maas recommends "power naps"—short breaks at the office in which you lay back and snooze instead of downing coffee or cola. But obviously this leaves out the vast majority of working mothers, who lack that kind of privacy.) Sapolsky warns against going for long stretches between meals when you're sleep-deprived because, given the elevated stress hormones you're already experiencing, your brain is receiving less glucose than normal. To avoid the roller-coaster effect of major fluctuations in blood sugar, he advises new parents to "go for the hunter-gatherer style of lots of little meals throughout the day."

The Power of Negative Thinking

So this is where you start: Your brain has been shrunken, marinated, and expanded. You've been jolted by trauma, and fried by lack of sleep. You've got a new brain, a Mommy Brain. But is it necessarily, even temporarily, a broken brain? This simply hasn't been proven.

The two studies that have made the strongest case for impairment were published in 1998 and 1999. In the first, researchers at Wayne University in Detroit led by Pamela Keenan compared pregnant women in their third trimester to a control group, and found the pregnant women forgot details of a passage read to them about 15 percent more of the time. (By three months postpartum, they were back on par with the rest of the group.) One year later, J. Galen Buckwalter, a University of Southern California psychologist, reported that when a group of pregnant medical students was tested for verbal memory (ability to recall lists of words) and learning ability, the women "just fell apart" in late pregnancy and up to two months' postpartum.

As other scientists have since pointed out, there were problems with both of these experiments. Each involved very small samples (just ten women in Keenan's study and nineteen in Buckwalter's), and the tests were never replicated. And Buckwalter didn't compare his pregnant volunteers with a control group, that is, nonpregnant women matched by factors including age and IQ. As Keenan acknowledged in an e-mail to me in 2003: "We do not have enough solid data to say without question that there is a memory deficit associated with pregnancy."

Moreover, in the years since the Buckwalter and Keenan experiments, three other larger studies from Australia and the United Kingdom have suggested that, as Helen Christensen says, "Pregnancy brain is a myth." Christensen, a cognitive psychologist at the Australian National University, acknowledges that, as an "aged mother" of three kids who found herself doing odd things such as storing soap powder in her refrigerator during her own pregnancies, she had a personal interest in the topic. Still, she had doubts about whether pregnancy itself causes brain drain. "I thought it could be because you're tired, you're sleep deprived and you're quite anxious about the upcoming event, but I wasn't convinced that it was really due to any brain damage," she has said.

In 1999, Christensen tested fifty-two pregnant women, with a thirty-five-member control group, on verbal memory, "working memory"—involved in learning, reasoning, and comprehending—and attention, while also investigating the subjects' moods. She found only one significant difference between the groups: The pregnant women were actually *better* at learning and remembering words that related to their condition. They perked up, for instance, when they heard the words "hospital," "placenta," and "labor." "It's like the cocktail-party phenomenon," Christensen says. "Even though there's a lot of noise, you can hear your name when someone mentions it across the room." A follow-up study by a colleague yielded similar results. Christensen boldly titled her published results "Pregnancy May Confer a Selective Cognitive Advantage."

Researchers at Charles Sturt University, also in Australia, echoed Christensen's conclusions after giving memory tests over the course of sixteen months to three dozen women, who were divided into groups of pregnant, postpartum, and control-group volunteers. In the journals they were instructed to keep, the two maternal groups reported that every day they seemed to be forgetting more things. One said she'd driven up to an intersection and suddenly found herself unable to recall whether a red light meant she should stop or go. Another said she had traveled more than a hundred miles on country roads to borrow a ladder from her sister, but then forgotten to bring the ladder back with her. Yet these women also performed no differently from the controls on objective tests. "Pregnant women and new mothers generally should be confident of performing to their normal cognitive capabilities," the researchers wrote.

Finally, in another small study published in 2003, scientists, led by psychologist Roz Crawley at the University of Sunderland in Great Britain, tested fifteen pregnant women on verbal memory, divided attention, and focused attention during pregnancy and postpartum, comparing them to controls. Again, they found the two groups did not differ in performance on objective tests, though the pregnant women believed themselves to be impaired. Crawley suggested that the women's strong perceptions of being overtaken by "Mommy Brain" might be a result of their own negative expectations of pregnancy's effect.

Here's where things become really interesting. Are you forgetting where you put the soap powder because your baby actually did eat your brain cells? Or might it be more that you've been trained to *expect* a decline, and when you do slip up, you are relieved to have a stereotype to explain it?

Today, with more mothers of young children than ever before involved in mentally demanding work, we have been given ideal circumstances in which to focus on possible lapses. At the same time, the Mommy Brain cliché suggests that the lapses we do detect are likely somehow related to our new reproductive status. In the Australian National University study, most of the pregnant women who were doing as well or better than their counterparts on cognitive tasks *thought* their memories were worse. As Christensen suggested, they would have small but dramatic forgetting episodes that they'd attribute to being pregnant, whereas other women might write it off as a routine glitch and forget about it.

The Australian psychologist Paul Casey, the leader of the Charles Sturt team, noted that several women in his study began complaining of memory problems from the time they became pregnant—despite Keenan and Buckwalter's indications that objective differences surfaced in the third trimester. What has truly changed, Casey believes, is the women's "metacognition": the way they perceive and judge their cognitive performance. In previous work, Casey has found that heightened self-awareness and reported forgetting tend to go together. He believes it's quite possible that pregnant women, known to be particularly immersed in their own experiences, simply remember all the times they forget things. "That," Casey says, "is a success of memory in itself."

Many pregnant women and new mothers are supported in their fear of a Mommy Brain drain by their friends, acquaintances, and culture. You might be a Nobel Laureate, but your kids' pediatric hygienist calls you "Mom." And magazines assume that your priority in life is to flatten your postpartum tummy.

"You're patronized constantly," says Laura Hilgers, a fellow freelance writer mother at my son's elementary school. "Just because I had a kid doesn't mean I had a lobotomy. But when you're out socially you discover that while, back when you were single, you could still be one of the boys, when you're a mom, you're relegated to the kitchen."

Mothers who start to anticipate these slights may take measures to defend themselves. The neuropsychologist Julie Suhr, at Ohio University, remembers when her first-born child, then three months old, developed viral meningitis. "I could tell that she was very ill, but she had no fever, so I knew I'd be viewed as a 'hysterical first-time mother,'" she says. "Since our pediatrician was in the same hospital that I worked in, I stopped at my office on the way to the appointment and grabbed my own white coat, with my nametag stating Dr. Suhr clearly displayed. I was clearly reacting to what I thought their expectations of me would be."

Suhr is actually an expert in expectations. Her research specialty is "stereotype threat," a term coined by Claude Steele, a Stanford University psychologist. Stereotype threat means that a member of a particular group, faced with a task thought to be poorly performed by members of that group, will *consequently* perform less well than they otherwise would on that task. Racial minorities, given expectations that they will test poorly on achievement tests, on average do so, as do women influenced by opinions that they will test poorly in math. Most relevant to the Mommy Brain stereotype is an experiment demonstrating that older adults subliminally primed with negative stereotypes about aging performed less well on tests of cognition than did other elderly subjects primed with positive images. Working mothers, burdened with negative stereotypes, may be similarly set up to fail.

Inspired by our conversation in 2004, Suhr set to work preparing an experiment to pinpoint whether new mothers' cognitive performance might be affected by the Mommy Brain cliché. "You'd be surprised at how little needs to be done to call up one's own negative biases," she says. Unrealistic expectations go a long way in this sense. For many new mothers, small errors are "unforgiveable," Suhr says, "whereas to others they are easily forgotten."

Counterattack of the Brainy Moms

In the 1980s, exasperated with the culture's acceptance of the notion of postpartum brain-death, a group of writers who also happened to be mothers began suggesting there might just be something positive about the way mothers used their heads. An important pioneer in this new movement was the philosopher Sara Ruddick, with her famous 1980 essay, later turned into a book, called *Maternal Thinking.*

Ruddick, then in her mid-forties, and with children aged eleven and thirteen, says she was writing in part as a response to feminists such as

Simone de Beauvoir and Betty Friedan, who suggested that motherhood would "ruin your life and be part of your oppression." Instead, she contemplated the idea that "there might be something of intellectual interest in mothering," an idea she found "strangely foreign and exciting." She was convinced that mothers develop a special, and in some ways enhanced, way of thought, as a product of their daily immersion in the concrete tasks of caregiving. In one memorable passage, she writes:

> Whatever mix of happiness and sorrow it brings, a commitment to fostering growth expands a mother's intellectual life. Routines of responsibility, exhausting work, and, for some, the narrowing perspectives of a particular profession or academic discipline conspire to undermine most mothers'—and most adults'—active mental life. But children are fascinating. Even as caring for children may reawaken a mother's childhood conflicts, in favorable circumstances her children's lively intellects rekindle her own. The work of fostering growth provokes or requires a welcoming response to change.

Carrying on with the brave notion that there might just be a brainy side to mothering, Camile Peri, in 1997 an editor at the Internet magazine *Salon,* helped launch a regular feature section called "Mothers Who Think" as a "place for mothers to exercise their brains." *Brain, Child: The Magazine for Thinking Mothers,* first published by two new mothers in Virginia, followed two years later. (Its title, according to Jennifer Niesslein, one of the editors, "is shorthand for something like, I have a brain; I have a child: don't condescend to me.") In 1998 at York University in Toronto, a group of women academics founded the Centre for Research on Mothering, its mandate being to promote "feminist maternal scholarship." *Literary Mama,* calling itself the first mother-focused online magazine about the complexities of motherhood, appeared in November 2003.

All these efforts have been steadily subverting the Hallmark card perspective of mothers surrounded by pink bows and pies. But even more promising in this regard is the legacy of the neuroscience breakthroughs of the 1990s, a scientific renaissance also known as the "Decade of the Brain." In that revolutionary time, researchers accumulated powerful evidence showing that the adult brain, which we once believed was set in stone shortly after adolescence, changes throughout life. We now know that people involved in specialized work physically change their brains as they dedicate themselves to their trades. The brain area devoted to a violinist's fingers, for instance, grows larger as he learns. Similarly, brain-imaging scans have shown that in London taxi-drivers, who must internalize maps of the city, the hippocampus—the brain's memory center—is larger than that of people who don't face such demands. You aren't really what you eat, but what you *do*, and mothers are no different. While scientists have yet to establish this, some suspect that repetitive *emotional* responses, including emotionally intelligent ones such as self-restraint or empathy, can strengthen neural circuits in the brain, making them more responsive—"like taking a walk through the woods time after time, and making a clearer trail," as one researcher described it.

This basic scenario of brain changes by means of experience throughout adulthood—referred to as "plasticity"—is one of the four pillars of this book, supporting the idea that motherhood can make women smarter. We now know that not only is the brain altered with new experience, but that positive, emotionally charged, and challenging experience can improve and help preserve its functioning, a phenomenon known as "enrichment." As I later describe in detail, children, when they're not driving you nuts, can often provide just that type of experience.

The second pillar has to do with the unique nature of a woman's bond with her child. Human childhood lasts longer than in any other species on Earth—and necessarily so does active and engaged human caregiving. Fortified by powerful hormones and secured with rigid cultural conventions, a mother's attachment to her offspring is usually stronger than any other relationship she'll ever have. Unlike marriages, friendships, or professional collaborations, rearing children comes with a profound obligation to keep working at challenges that might otherwise be abandoned—challenges, moreover, that constantly change and become more complex over time.

Because most of these challenges engaging mothers' brains are inherently emotional, they've long been undervalued as a route to greater intelligence. Yet our culture's conception of what constitutes intelligence has been changing over the past two decades as scientists have gained a new appreciation of the importance of the brain's emotional networks in effective decisionmaking. That emotional skills are a key part of being "smart" is the third pillar, or main argument, of this book.

These three ideas deal with *how* mothers become smarter. The fourth pillar addresses the *why,* which is basically a matter of evolution. Like every other living being, we're driven to do our best so that our genes outlive our bodies. "A chicken is only an egg's way of making another egg," as Samuel Butler famously observed. A female's evolutionary importance peaks when she is picking a mate, bearing her children, and helping them survive to maturity. We may be doing our mothering, these days, in between performing brain surgery or addressing the United Nations, but in the overall context of our several million years of evolution from nonhuman primates to humans—during which the brains we have today took shape—there is no other time when we need to be so smart. Only recently have scientists started to speak of a distinctly "maternal brain." Now, for the first time, powerful new technology is helping to pinpoint ways in which this new brain may be an enhanced brain.

CHAPTER
3

The Nearly Uncharted Wilderness of Mothers' Brains

She (woman) has a head almost too small for intellect but just big enough for love.

1848 OBSTETRICS TEXT

INSIDE THE HUGE metal cylinder of the MRI machine, Tara Magnuson lies still as she listens to a tape recording of the cries of her four-week-old baby, Alexander. Magnuson knows that Alexander is sleeping calmly in his stroller just outside the door, snugly wrapped in his blue blanket and guarded by his father. Even so, she breathes a bit more quickly, anxiety flickering through her mind.

A few feet away, on the other side of a large window in the Yale University basement, a computer is storing images of Magnuson's brain. Splotches of red and orange, marking flowing tides of blood, track the new mother's changing emotions.

With the help of Magnuson and other paid volunteers, psychiatrists James Leckman and James Swain are studying healthy parental concern to understand better what goes wrong when it's absent or out of whack. They also share an interest in obsessive-compulsive disorder, Leckman's hypothesis being that this illness—with its trademark slavish checking, rituals, and fear of germs—springs from the same evolutionary origins and employs the same brain circuits as normal parental behavior. Yet as the father of two grown children, Leckman in particular has long been fascinated by the broader question of what might be happening in a parent's brain that supports the everyday miracle of devoted childcare. What supernatural force could make an otherwise reasonable adult invest such energy in coaxing along a life form that does little during its first several weeks on Earth but wail, poop, and eat?

"Parenting is a complete transformation of your hedonic homeostasis," he theorizes as he sits in the cold autumn light of the old stone-walled Yale library courtyard. In other words, a parent's personal sense of comfort discards old standards for new ones, these encompassing the welfare of another human being.

Leckman was taken aback when this happened to him. A couple of decades ago, in the midst of the heavy demands of his medical internship, he found himself building a cradle from scratch for his yet-unborn daughter. At the same time, he felt suddenly plagued by worries about possible disasters—infections, birth defects, congenital abnormalities. As Leckman observed these changes in himself, he noticed that his wife, Hannah, a graduate student in French literature, was undergoing a similar transformation. Both parents-to-be found their thoughts constantly turning to their daughter-to-be, just as if they were beginning a romantic infatuation.

Eventually, Leckman determined to investigate this most ancient experience with one of the most modern scientific tools available: brain-scans that let scientists track emotions as they ripple through the mind. This revolutionary technique is just beginning to help illuminate the powerful workings of the 300 million feet of wiring, packed into a 1 1/2-quart space, that constitutes a mother's brain.

Moving Pictures of the Mind

Once used purely for medical purposes, MRI brain scans by the mid-1990s were increasingly supporting scientific research. Functional magnetic resonance imaging (*f*MRI), which records brain activity as it is happening, has been a breakthrough in this regard, illuminating the cerebral processes behind a myriad of emotional states, among them gambling, cravings for chocolate, choosing a car, winning the lottery, and having an orgasm (for that last one, unsurprisingly, we can thank researchers in the Netherlands). By the first years of the new millennium, the data were coming in at a furious pace and beginning to clarify the mechanisms behind such fundamental but scientifically understudied emotions as love, empathy, and parental affection.

Functional MRIs rely on the unique magnetic properties of blood, which contains iron. When an area of the brain is being used, such as when you think, or feel an emotion, blood flows to that region, bringing with it the energy needed to accomplish the task; in the process, the tissue's magnetic properties change. By exposing the brain to a powerful magnetic field, the *f*MRI scan can take a series of pictures, much like making a cartoon, that capture the changing flow of blood. Images emerge of the brain in motion, the hard-working regions "lit up" in bright colors.

Today, researchers in at least half a dozen labs in the United States, Britain, and Switzerland are using *f*MRIs specifically to study the brains of mothers, and occasionally also fathers. Like Leckman and Swain, they use tape-recorded baby cries and sometimes endearing baby photos to stimulate the neurobiology of parental love. These scientists all are focused, as Leckman comprehensively describes it, on "the interaction of fearfulness, vigilance, reciprocal selective recognition and reward that resonates within a place of wonder that allows successful parents to see the world from a new vantage point and invest in their child."

Leckman and Swain suspect the scans of the new mothers will eventually reveal some similarities with past imaging of brains of people suffering obsessive-compulsive disorder. If so, they'll show more activity in the right side of the frontal cortex, a brain area key in detecting threats. At the very least, says Leckman, news of such a similarity could increase our compassion for people who suffer from obsessive compulsive disorder (OCD) by demonstrating that "we really all have something in common." (In preliminary findings, at this writing, Leckman and Swain have seen diminished activity in part of the amygdala, the brain center of fear and gut feelings, among their volunteer subjects in scanning intervals from two weeks to three months postpartum. In that same period, the volunteers have reported a diminishing amount of OCD-like behaviors. Leckman is still trying to learn more about this possible correlation.)

Why Moms?

The Yale study is just one indication of scientists' increasing eagerness, over the past couple of decades, to probe mothers' minds through research on rats, monkeys, and humans. Neuroscientists, psychiatrists, psychologists, and sociologists have joined in the belief that the parenting side of the parent-child relationship, so key to the survival of the species, is worth studying in depth. As Michael Numan, a Boston College neuroscientist, wrote in the 2003 book he co-authored with Thomas Insel, *The Neurobiology of Parental Behavior*: "We are . . . delving into neurobiological factors that may have an impact on core human characteristics involved in sociality, social attachment, nurturing behavior, and love. In this very violent world, it is hard to conceive of a group of characteristics that is more worthy of study."

The new wave of research is actually a niche in a recent larger trend in which scientists have flocked to the study of "positive" emotions, such as love. Some attribute this shift to a changed zeitgeist, especially after the terrorist attacks of September 11, 2001, which brought home the danger of a world in which hate trumps love, and seemed to raise the value of skills we've come to think of as "emotional intelligence." One year after 9/11, a philanthropic enterprise called the Institute for Research on Unlimited Love made its first nearly $2 million in grants to scientists, including Leckman and Swain, studying virtuous emotions. Yet even by the late 1990s, neuroscientists were acknowledging that although we knew a great deal about depression, anxiety, rage, and drug abuse in humans, we knew much less about the mental engines of empathy, social bonds, altruism, and joy. And when you're looking for a good place to study empathy, social bonds, altruism, and joy, where better to start than a healthy mother looking at her newborn baby?

"Most of the time, people come out of the MRIs looking really fatigued, because they have to stay there for up to two hours and keep really still; they can't move their heads a centimeter," says Jack Nitschke, a neuroscientist at the University of Wisconsin, who has scanned mothers looking at pictures of their babies, seeking clues about the brain mechanics of happiness. "But when the moms come out . . . they are radiant, saying how great it was."

A landmark event in the changing focus of parenting research was a 1968 paper by the psychologist Richard Q. Bell, who claimed that offspring had as much or more influence in socializing their parents as the other way around. Several more articles and books in a similar vein pointed out ways in which the child, previously seen as much more passive, was wielding influence over his parents. Mothers tend to look in the same direction as their babies more often than infants follow their mothers' gaze. And nine times out of ten, babies will be the first to make or break eye contact with their parents.

It's probably no coincidence that these ways of thinking about parenting have been emerging at a time when more women have been coming into prominence in science, and more men are acting as caregivers at home. Consequently, many more engaged parents have been filling the ranks of scientific opinion makers. "Research is Me-search," the saying goes. Driving many of the most accomplished scientists is the quest to understand problems relevant to their own lives.

Take Tracey Shors, a Rutgers behavioral neuroscientist, and new mother, who has been studying the effects of stress on mother rats. Shors is particularly interested in postpartum depression, even though she describes her own experience of giving birth at age forty-two in glowing terms. She delved into parenting studies only after she had acquired tenure because she was afraid the subject might be perceived as "soft" science.

Another tenured mother, C. Sue Carter, a University of Illinois behavioral neuroscientist, has become a leading expert on oxytocin—a hormone essential to labor and lactation. She dates her interest to the birth of her first child, when her obstetrician gave her an extra dose of synthetic oxytocin to speed up delivery.

Similarly, Kerstin Uvnas-Moberg, a Swedish neuroendocrinologist, switched from studying gastric juices to focusing on oxytocin after experiencing "a systematic change in [her] behavior and way of thinking" in the process of mothering four children. And Alison Fleming, a psychologist at the University of Toronto and the mother of three girls, has accomplished pathbreaking work in exploring the development of maternal motivation and behavior—work inspired by her childhood as the daughter of career-focused parents who sent her away to boarding schools.

Until recently, women were vastly outnumbered not only as prominent scientific researchers but also as subjects. Male scientists determinedly preferred to study what they called the *normal brain* versus the more variable *cycling brain* of females whose hormone levels were constantly changing due to their menstrual rhythms. In the field of stress physiology, for instance, as late as the early 1990s, U.S. women made up only about 17 percent of participants in studies—despite evidence that women were even more prone than men to stress-related disease. But in 1995, the U.S. federal government, heavily lobbied by women's groups and female scientists, required that new research projects had to include members of both sexes, a change that helped open the way for a new level of understanding of female and maternal health.

Maternal Riches

To realize how much times have changed, consider the story of Marian Diamond, a mother of four and so respected a neuroanatomist that, in the 1980s, she was one of the few scientists in the world permitted to examine Einstein's brain. Several years earlier, Diamond had made a dramatic discovery about mothers' brains that was all but ignored by her colleagues, and even today remains surprisingly obscure.

"I was the only woman grad student in my discipline from 1948 to '52," she recalls. "Now, 50 percent are women. The male attitude was that I should have been home taking care of my children. I don't blame them, of course. It was just their hormonal orientation."

Diamond acknowledges that she was following her own hormonal orientation at the time. She gave birth to her first child at the age of twenty-six, after which she shifted to part-time teaching and research for several years so that she could volunteer at her children's preschools and be at home in the afternoons. "When I held my first baby in my arms, my hypothalamus told me: *This is why you are here,*" she says, referring to a part of the brain that secretes hormones that induce maternal behavior.

In the 1960s, Diamond joined a University of California at Berkeley team that was making exciting discoveries about the impact of "enrichment"—stimulating experience—on the brain. The researchers had demonstrated that if a caged rat's environment is made more interesting by the presence of toys or other rats, it can lead to growth in the rat's cerebral cortex, and, in turn, to better performance in navigating mazes.

To keep things simple, these scientists routinely worked with male rats, but Diamond chose to work with females. In one of her projects, she set out to determine whether female rats with enriched brains passed that benefit down to their pups. To do so, she had to first perform autopsies to confirm that an enrichment effect had taken place in the new mother's brain. But when she looked at the rat mothers' brains, Diamond was surprised to find no difference between those that were enriched and those that were impoverished. Could it be, she worried, that the female brains simply didn't respond to enrichment? Yet when she compared impoverished and enriched virgin females, she immediately saw the difference in the thickness of the cerebral cortex. She realized that for the mentally impoverished female rats, *pregnancy* had been an enrichment. "That made sense to me," she recalled. "Pregnancy has to prime you to take on responsibility for the survival of a new individual, to give you what it takes to look after the little one." In humans, too, Diamond contends, motherhood can be "phenomenal enrichment for the brain."

Diamond published a paper on this finding in 1971, but says that "no one paid any attention to it." So despite her continuing curiosity, she never followed up that line of research. "It was only much later, when other women succeeded in their fields," she says, "that my contributions were also recognized."

The Minefield of Motherhood Research

Indeed, throughout the 1960s–1980s, politics put a damper on scientific research that called attention to gender-based brain differences. And this certainly included studies on the influence of reproduction on women. Feminist scholars in sociology, psychology, and social anthropology seemed to be on the lookout for suggestions that women were innately oriented toward caregiving. They feared, with some good reason, that it would encourage pigeon-holing them as mindless nurturers, just as professional women in hitherto male-dominated fields were testing their wings. As Susan Faludi wrote in her bestseller, *Backlash,* "'Difference' became the new magic word uttered to defuse the feminist campaign for equality."

Sara Ruddick, the author of *Maternal Thinking,* recalls that she was fiercely criticized for being so "essentialist"—that is, overemphasizing the difference between women and men—as to suggest that mothers might see the world differently, even though Ruddick insists it's the *work* of caregiving, more than innate differences, that leads to such a change. The sociologist Alice Rossi, a prominent sociologist and founder of the National Organization for Women, says that in 1983, when she gave a talk on what she describes as "a biosocial perspective of parenting," she caught similar flak: "I was suggesting there were genetic differences between the sexes that were important for the primary roles of males and females, and that social change efforts to bring about sex equality would not succeed on the premise that sex differences were only a matter of differential socialization," she recalls. "Rather, males would need compensatory training in nurturing skills, females in assertiveness skills. But that was something feminists didn't want to hear. They were still arguing . . . that any differences we see between the genders are because society has forced women into those roles."

These perceptions were clearly waning by 1993, however, according to Mark George, a physician specializing in psychiatry and neurology at the Medical University of South Carolina. George that year was working at the National Institutes of Health (NIH), brain-scanning men and women as they remembered sad events in their lives. Women, he found, activated many more brain regions than men when reliving those sad moments, perhaps a clue to understanding why the rate of depression in women is twice as high as in men. "They lit up like a starry sky," as he describes it. As much as he recognized the significance of this discovery, George initially worried that it might be "taboo" to focus on such a fundamental difference. "I presented the results to a meeting of scientists, and they said go ahead and publish," he recalls. "In the time it took for me to do so, the topic went from totally taboo to trendy. It was amazing to see how society changed." That change has freed researchers to follow their hunches and pursue a deeper understanding of the transformational mechanics involved in becoming a parent.

The Evolving Mommy Brain

While Tara Magnuson, the Yale brain-scan volunteer, prepares for her *f*MRI exam, James Swain runs through a list of questions. Does she wear braces on her teeth? Does she have shrapnel injuries? A metal plate in her skull? A pacemaker? Magnuson listens, shaking her head "no" to each question, her face bearing the fixed smile of the chronically sleepless. As she does, she deftly prepares Alexander for her absence. Keeping her eyes on Swain to help her concentrate on his questions, she positions her infant under her blouse, nurses him, burps him, and changes him from one side to another. Alexander's impatient bleats soon turn to contented coos.

In this efficient response to the varied demands of the moment—Swain's need for accuracy, her baby's for food, affection, and proper digestion, and her own modest desire to avoid flashing her breast—Magnuson seems like a minor icon for the theories of Paul MacLean, the visionary NIH psychiatrist who is in fact indirectly responsible for the scene. MacLean has spent much of his career crafting a consummately positive portrait of the Mommy Brain, emphasizing its unique evolutionary impact on social bonds and language. It was his idea to scan mothers' brains in the first place.

"For more than 180 million years, the female has played the central role in mammalian evolution," MacLean has written. Maternal behavior, he theorizes, was what first began to separate mammals from reptiles, some 250 million years ago. Lizard maternal care amounts to laying eggs and slithering away. But the mammals that sprang from those deadbeat moms now share three essential behaviors, all directly derived from the mother-child bond. They use language, they nurse their young, and they play, a behavior that may have originated from a family's need for peace among siblings competing in the nest.

MacLean's most famous theory is that the human brain physically developed, over time, together with this behavioral change. He introduced the concept of the "triune," or three-in-one brain, an integrated system governing the body's basic functions, emotions, and thought. The oldest, most basic and ensconced part of this triad—known as the "reptilian brain"—is the brain stem, striatum, and parts of the thalamus. The brainstem controls involuntary functions: breathing, heart rate, reflexes. The striatum governs motivation to do things and automatic habits, such as when you unconsciously pick up your keys before leaving the house. Parts of the thalamus are important in processing sensory information, such as when you touch a hot stove.

Next to evolve, and sitting on top the "reptilian brain," was the "emotional brain," or limbic system. It includes the almond-sized amygdala, responsible for fear and other "gut" reactions important in self-preservation; the hypothalamus—the hormone secretion center; and the seahorse-shaped hippocampus (the word comes from the Greek *hippo,* horse, and *kampos,* monster) which, as mentioned earlier, is in charge of memory and learning.

Finally, there's the "new bark" of the neocortex, the outermost and newest part of the brain. Present in all mammals, but largest in humans, the neocortex (also referred to as the cortex) is responsible for critical thinking, flexibility, language, and long-term planning. Throughout several million years of prehuman and human history, the vast bulk of which we lived with no idea at all about physics or French literature, the main purpose of the neocortex was to help mammals manage their social relationships, including parent-child bonds and ties with other adults who help improve children's chances of survival.

Now in his nineties, MacLean has many active disciples, including Jeffrey Lorberbaum, the brain-scanner in South Carolina. As a medical student, Lorberbaum developed a passionate interest in MacLean's evolutionary approach to behavior, and in 1997, while interviewing for a postdoctorate fellowship position with Mark George, found that George shared his enthusiasm. George knew MacLean personally, and mentioned that he had been urging him for several years to brain-scan mothers in hope of zeroing in on the brain regions responsible for maternal behavior. Lorberbaum jumped at the chance to devise the first such experiment. He agreed with MacLean that a mother's response to her baby's cry could provide a vivid moving picture of a brain engaged in parenting.

From Pecans to Cantaloupes

Several years before Lorberbaum started brain-scanning human mothers, the Boston College neuroscientist Michael Numan, in a series of careful experiments, had mapped out the rat's maternal circuitry—a neural system connecting active parts of the brain when a rodent responds to her pups. The hub of this neural path is a structure called the medial preoptic area (MPOA), located in the front part of the hypothalamus, the hormone center. Numan's research indicates there are neurons in the MPOA that become more active with the hormones of pregnancy, a reaction he thinks underlies a female's nurturing behavior. He found that these neurons are connected to midbrain areas that are important in feelings of motivation and reward, including the ventral tegmental area (VTA) and nucleus acumbens (NA), the latter often called the brain's pleasure center. This separate "reward circuit" activates upon winning money, looking at a handsome face or even being high on cocaine, when the neurotransmitter dopamine travels along it, carrying a feeling of delight.

By 2004, Lorberbaum had brain-scanned more than forty mothers and ten fathers and had found strikingly similar results. As he analyzed his data, showing the MPOA, the VTA, and the NA all lighting up when mothers heard their babies cry, Lorberbaum felt awed by the apparent connection we share with the rest of the animal world. "You could see the brain wasn't reinvented just for humans," he says. He was also buoyed by the notion that the large body of literature available on animal mothers might now be seen as more useful in throwing light on human behavior.

Scientists have long recognized that rodents have a lot to teach us. The pecan-sized brain of the rat is remarkably similar in composition, and is governed by the same neurochemicals, as the cantaloupe-sized brain of a human. Indeed, the great bulk of what we know to date about the human nervous system and its links with reproductive hormones has been learned from studying animals. This is so in part because it has been much easier to "sacrifice" animals for the purpose of peering inside their brains than it would be with humans. But just as important is that animal experiments can be so easily controlled, factors that would irredeemably complicate human trials being ruled out.

Imagine the difficulty of trying to compare the mental attributes of two women of similar age—one a mother, the other without children—and teasing out the influence of parenthood? Even if their IQs and income levels were the same, so much else, from heredity to diet to outside interests, might be different. This helps explain why so few researchers to date have ventured to try to understand how humans change when they become parents. Still, there have been some interesting recent exceptions.

In 2002, James J. Dillon, a psychologist at the State University of West Georgia, surveyed thirty-five parents and fifteen teachers, asking them to discuss occasions when they had learned something valuable from a child, or when a child had changed them in some significant way. Not one of the participants, split equally along gender lines, said they had never experienced such a change. Among the parents, the greatest percentage—34 percent—said that children had been a catalyst for them to better understand themselves. Twenty-nine percent cited what Dillon called *inspirational influence*, and some 26 percent said their kids had affected their attitudes and behavior by bringing them important information or ideas, such as the need to care for the natural environment.

In researching this book, I persuaded two scientists to perform custom analyses of data they had already collected. In 2004, Lorberbaum and the Turkish psychiatrist Samet Kose surveyed thirty women whom they had already brain-scanned, asking what changes, if any, they had undergone in the first two months after delivering a child. A high percentage said they had become warmer and kinder in general and that they understood others' points of view more easily, were less anxious about the way they appeared to others, could handle stress better, and could multitask—and find things in a grocery store—more skillfully.

The second scientist, Ravenna Helson, a research psychologist at the University of California at Berkeley, called on a much larger sample. For the past forty years, she and her colleagues have been following 121 graduates of Mills College, an elite women's school in northern California. At my request, also in 2004, she compared mothers and nonmothers, synthesizing findings from three personality inventory studies performed over five years between the late 1950s and early 1960s. Helson found that women in the first and larger group showed "a more significant increase in psychological understanding of self and others and a distinctive increase in responsibility."

Still, Helson warned that women shouldn't assume that having babies will necessarily change them in these ways. The Mills graduate mothers were all products of their temperaments, their times, and their choices—as are women today. "What happens to the mother," Helson stresses, "depends on her personality and goals and the context in which she mothers."

Love Junkies

The tremendous variability among human mothers naturally makes us more unpredictable than rats. But one thing so many of us seem to share with the rest of the animal world is that deep feeling of connectedness that keeps us automatically moving *toward* the objectively annoying sound of our babies' cries, rather than running out the door. In this way, women may change their behavior almost as dramatically as do female rats, which, before they become mothers, behave fearfully toward rat pups, reacting as if they smelled bad and occasionally trying to bury them. Lorberbaum believes the change has something to do with the intersection between what he calls the "maternal circuit" and the "reward circuit" that he noticed in his brain scans when women listened to their babies.

Interestingly, and perhaps as many new mothers may have suspected, fathers appear to have a very different basic reaction to their own babies' cries, according to preliminary results of research by Lorberbaum and Kose, which were still unpublished as of this writing. While mothers reacted with the ancient, emotional reward centers, fathers responded only with the newer-evolved parts of the cerebral cortex involved in thinking and planning. Lorberbaum speculates that this is because intensive fathering is a relatively recent evolutionary development, while mothers, over many generations, have become more hard-wired to care deeply, and as he says, may help explain "why mothers are the first to jump out of bed."

It remains to be proven whether or not mothers listening to their babies' yowls experience a dopamine surge, but the idea makes sense when one considers the way Donald Symons, an anthropologist at the University of California at Santa Barbara, has explained the evolution of emotions. Symons speculates that our most powerful urges have grown to meet the basic survival needs that are most difficult to achieve. We are driven to seek food and love, scarce as they may be, and to care for our children, annoying as they may be, and to feel good about it all, thanks to this crafty reward system in our brains.

Indeed, as researchers explore these ancient motivations with *f*MRIs, they are finding basic similarities between the love that women feel for their romantic mates and for their children, which in turn is similar to the passion that addicts acquire for their drug of choice. At the same time that Lorberbaum was publishing his initial studies, another brain-scanning group, at University College, London, reported that the reward circuit activates when women look at pictures of their lovers, as opposed to other people they know. Then, in 2004, the British researchers saw a similar result when they scanned twenty mothers gazing at photographs of their own children, aged from nine months to six years, compared to similarly aged children of acquaintances.

Now, most new mothers don't need to have their brains scanned to know that baby love is like a drug—addictive and mind-altering. "We were looking at old baby pictures the other day, remembering how absolutely perfect and gorgeous he was, and there he was, so cross-eyed and funny-looking!" reports one mother I know. Rat mothers seem to share this blinding early infatuation. In one study, new rat mothers consistently preferred to enter a room in which they had been exposed to pups instead of a room in which they had been exposed to cocaine. (After about two weeks, the moms preferred looking for the drug.)

Joan Morrell, a neuroscientist at Rutgers University, who designed this study, thinks it helps explain why human women hooked on cocaine more often choose detox treatments while pregnant or shortly after giving birth—just as many pregnant women give up coffee and cigarettes. There's something about the hormones of pregnancy and early motherhood, she suspects, that gives women the extra motivation to kick their habits. "Pregnancy and the nursing period are unique in the natural history of the human, or of any mammal, because during this period the child can be uniquely harmed or uniquely helped, each with life-long consequences," she writes.

The Maze of Motherhood

If, as Jeffrey Lorberbaum's brain scans suggest, human mothers have more in common with rats than we may at first want to believe, it's encouraging to consider the work of Craig Kinsley and Kelly Lambert in

Virginia. They're the scientists who've found a conspicuous maternal advantage in the learning and memory capacities of rats tested on their performance in running through mazes.

Kinsley and Lambert began their experiments in 1996, before any human mothers had been brain-scanned, and they appear to be the first neuroscientists on record to suggest that parenting can lead to concrete brain improvements. "I was amazed when I first got into this, because what could be more significant than giving birth?" says Lambert. "The fact that something can be such a huge life event yet ignored for so many years is shocking to me."

Lambert, a tall, attractive blonde who drives a bright red truck, defies the traditional stereotypes of neuroscientists, not least in that she devotes herself to her high-pressure job while also raising two preteen girls. As chair of her college's psychology department, her normal schedule, when she isn't traveling, is to teach and conduct research all day, return home for dinner, put the children to bed at 8:30 P.M., and work on writing projects until past midnight.

Lambert grew up poor in Mobile, Alabama, and used to search for toads in the woods near her home. Her mother told her not to touch them, warning that they could give her warts, but, she says, "I'd let them pee on me just to show her that wasn't true." She shares this investigative fervor with Kinsley, who as a child would pick fleas off of his cocker spaniel so that he could observe them feeding on his arm. They have been collaborating since the early 1990s, when Kinsley decided to study maternal behavior after observing the first pregnancy of his wife, Nancy. "I saw her change from someone who lacked confidence and was apprehensive and ambivalent to someone with great competence and ardor," he says.

Supermoms

In the first years of the new millennium, several studies were published revealing surprising ways that children begin influencing their mothers—and, in some ways, fortifying them—as early as conception. One report showed that women pregnant with boys tend to eat about 10 percent more calories a day than those carrying girls, yet don't gain more weight. Another found that mothers who breastfeed heal faster when wounded. Although both of these effects are temporary, researchers looking at a range of mammals were also finding signs of some permanent change. Second-time mothers, for instance, react faster and more efficiently to their babies, a crude but striking way of showing that some long-term change has been made. Second-time mothers throughout mammal species also tend to have more milk and lower levels of stress hormones than first-timers. This would underpin something many mothers already suspect: that the birth of their first child can be a kind of rebirth for themselves.

"The brain is changed when an animal learns," says Fleming, the University of Toronto neuroscientist, who has studied rodents, sheep, and humans. The learning she refers to is much more profound than anything you'd pick up in your Lamaze classes; it's a process involving your most basic biology, and it makes significant impacts on your brain, your hormones, and your behavior far into the future.

One of the most remarkable changes evoked by parenting was revealed by brain-scanning researchers at the University of Basel in Switzerland in 2003. Comparing volunteers who listened to recorded baby cries and laughter, they showed that in mothers and fathers, the amygdala and interconnected regions important in emotion activated more strongly in response to the wails than to the chuckles, whereas nonparents' brains activated more in response to laughter. As Erich Seifritz, the research leader, told reporters, this makes biological sense. Tears are a signal that something is wrong and that you, the parent, must fix it, if you want your genetic investment to pay off. The fact that a nonparent is more responsive to the good-time Charlie soundtrack shows that the parent's less fun but necessary response is a learned response. And remember: Learning physically changes the brain.

Does this basic urge to help your crying baby constitute a form of smartness? If you believe in evolutionary theory, the question is a no-brainer: There is nothing more important for you to do, and to do well.

In the Pleistocene epoch, the hunter-gatherer time between 1.6 million and 10,000 years ago, during which human brains and behavior were basically shaped, children faced perilous odds against their survival. Because men were usually out looking for meat, the women, mostly, had to stretch their capacities to care not only for themselves but for the new people in their lives. And as human childhood has grown steadily longer, and human culture more sophisticated, this caregiving job has required an increasingly sophisticated mental toolbox. As the anthropologist Sarah Blaffer Hrdy has written, any woman "who relied on looks alone to pull offspring through was not likely to be a mother very long or leave descendants."

That simple goal of pulling offspring through requires a range of mental skills that can be useful both within and outside of one's own family—including those five key attributes of perceptiveness, efficiency, resilience, motivation, and sociability. The next chapters will describe what scientists know and are learning about these potentially enhanced aspects of the Mommy Brain.

Part Two

THE FIVE ATTRIBUTES OF A BABY-BOOSTED BRAIN

CHAPTER

4

Perception

The Expanding Realm of a Mother's Senses

The smell and taste of things remain poised a long time, like souls, ready to remind us . . . and bear unfaltering, in the tiny and almost impalpable drop of their essence, the vast structure of recollection.

MARCEL PROUST,
REMEMBRANCE OF THINGS PAST

"CHECK THIS OUT," says Craig Kinsley, as the image of an albino rat flickers onto the projector screen at a University of Richmond conference room. The rat, a new mom who has just nursed her pups, sits calmly in her cage while a live cricket is dropped in a few inches away. The rat's nose twitches once, as she senses the insect's presence, and in the next instant, she leaps upon the cricket and gobbles it down.

In a similar scene played moments before, a virgin rat had taken fully three times as long to catch the cricket; in fact, she'd seemed oblivious to the insect's presence as it recklessly explored her cage. This same disparity had shown up repeatedly in tests of rat mothers versus nonmothers. Kinsley, standing next to the screen, watches with fatherly pride. He refers to this experiment as "mom rats kicking virgin butt."

It seems clear that this competitive edge comes from some strengthened ability, but Kinsley is still trying to figure out what that ability might be. It may just be a matter of motivation: Although both groups of rats had been kept just as long without eating, the nursing moms were probably keener to store up calories. They were looking out not just for

Number 1 but also for Numbers 2, 3, 4, and so forth. But Kinsley suspects his butt-kicking rats are also gifted with some greater sensory power, sharpened by evolutionary necessity. "Whatever system you're looking at, if it enhances the possibility that pups will survive, it's likely that it is improved in the moms," Kinsley believes.

The same basic rule probably also applies to humans. Even though pregnancy and the early postpartum weeks are often seen as a time of torpor, many women say they actually feel more sensually alive, more vibrantly responsive to their babies and to the world in general. "It's the only thing that ever beat taking mushrooms in Machu Picchu," a Santa Cruz social worker friend once memorably said.

You may smell and taste things differently when you are pregnant; you may crave some foods and feel repulsed by others. Once your baby is born, you may find yourself waking up at the same instant as he does, or even a few seconds before. You may quickly learn to recognize his scent and his cry, and understand whether that cry means hunger or pain, boredom or fatigue. And, though haunted nonstop by visions of disaster, you will somehow avoid clunking his head against the banister while carrying him and manage to catch him each time he nearly rolls off the changing table. All these reactions reflect subtle changes in your brain—some temporary, some longer term.

These sharpened perceptions, enlivening a mind swamped with dread, angst, moments of glee, and a fierce desire for sleep, may be at the root of what we mean when we talk about a mother's instinct—not the instinct to *be* a mother, but what happens when there's no turning back: the way you know immediately if your child is sick, in danger, or just trying to get away with something. This is more than a matter of any of your five senses improving: You're paying attention, and quickly learning from experience, because someone's life depends on it.

Nasal Instinct

It starts with the nose, so basic a conduit for our feelings that the emotional, "limbic" brain areas have also been called the *rhinencephalon,* or "smell brain." Marcel Proust recognized the potent interplay between smell, taste, emotion, and memory a century ago when he famously wrote of the memories that came flooding back with the bite of a petite Madeleine cookie. Today, we know that our olfactory bulbs, where smell is first perceived, have an ancient link to the amygdala, the brain center for gut feelings, thus giving odors a subterranean power to record and trigger memories. Studies on animals and humans show that this deep and mysterious connection is a particularly important compass for us as we navigate our journey through motherhood.

Kinsley suspects a heightened sense of smell or vision in early motherhood may best explain the mother rats' superior performance on the cricket test. He has ruled out changes in hearing through an experiment in which the cricket's sounds as it moves through the rat's cage are covered by white noise. Recent research also offers compelling reason to believe that a mammal female's sense of smell is somehow changed in the course of pregnancy and delivery, made temporarily more sensitive and discriminating—or, if you will, smarter.

Clinical scientists at the University of Calgary in Canada discovered in experiments on mice that, in early pregnancy, a huge burst of incipient neurons—stem cells—are produced in the forebrain. These cells then migrate to the olfactory bulbs, the brain structures that interpret smells, and come online with delivery. That's obviously a key time for the mother mouse and her baby: the first encounter that determines the strength of their future bond. The quality of those first sniffs can determine the odds of a baby's survival. And that's precisely why mice mothers acquire what appears to be a strengthened smelling capacity at that moment, the Canadian researchers believe. They've concluded that prolactin, a hormone that surges when the mouse mates, is responsible for the change. And they think the same thing may happen in humans, who also experience surges in prolactin when *they* mate.

Samuel Weiss, a neurobiologist and one of the team's members, calls the discovery "the first example of how a body reacts to a physiological phenomenon, which we think is linked to new brain cells, which are linked to a new behavior." Only relatively recently has it been known that the adult brain forms new neurons all the time. Yet, until the Canadian finding, these neurons were assumed to be most important in helping to replace loss and repair damage. Now, Weiss said, a new role was clear: that of helping new babies survive.

This powerful role of smell in bonding may be much more important for mammals such as rodents than for humans, who are under strong cultural pressure to remain with and take care of their babies. (Sheep will neglect their newborns if prevented from sniffing them during the first hour after birth.) Yet smell is more important in humans than has long been assumed.

Until the 1990s, scientists routinely dismissed a major role for our noses in our emotional behavior, in part because the human olfactory bulbs are proportionately so much smaller than they are in other species. And common wisdom until about the middle 1990s was that humans lacked another small sensing device, prevalent in other mammals, called the *vomeronasal organ.* A tiny pair of pits positioned against the septum in each nostril, the vomeronasal organ is used for detecting chemical compounds in body odor, called *pheromones.* Today we understand that humans indeed have all the necessary parts to detect and interpret the subtlest messages we pick up with our noses, and that these cues can be fundamentally important when we choose a mate and care for our young. They often steer our behavior without registering in our conscious minds.

Pheromones, which have been called "biochemical bouquets," are believed to be the reason that women who live in close proximity to one another, as in college dorms, find that their menstrual cycles converge. (Scientist wags call this "unicycling.") When it comes to dating, body odor—or "chemistry"—turns out to be critically important in the search for a suitable mate. In what may be an ancient mechanism that helps dissuade mammals from reproducing with their own kin, ovulating human females, and mice, prefer the scent of males with different immunological profiles than their own. Mice pick up these cues from urine. Women sniff sweat. Tests have shown that women can distinguish between T-shirts worn by different men, and can rank them as seductive or not, the results showing up as genetically *dissimilar* or not. Women also tend to have a particularly heightened sense of smell at ovulation, when it's easiest to become pregnant.

During pregnancy, a female's sense of smell starts to play another vital role. Research has shown that, just as with a woman who is ovulating, a pregnant woman becomes more sensitive in her detection and identification of odors. Moreover, many smells that she easily tolerated in the past become suddenly impossible to bear. "I remember walking by a garbage dump with a friend of mine who was pregnant, and she was yelling for us to get out of there, while I hardly noticed it," recounts Karen Parker, a Stanford University psychology postdoctoral fellow.

This heightened sensitivity is useful during pregnancy, when you're sharing every meal with a vulnerable embryo. Because foods you should avoid could be more likely to carry parasites or pathogens, you are given help in avoiding them through your reaction to certain odors. These odors may even make you vomit. Food aversions, which begin just a few weeks after conception, when the baby's organs are just starting to form, are usually at their strongest when they're most needed. During this time, a woman's immunological defenses are reduced, making it less likely that she'll reject the "foreign body" of her new baby, but leaving her temporarily more vulnerable to disease.

The notion that food aversions, and even morning sickness, may be signs of the body's wisdom has been bolstered by research since the late 1980s. Margie Profet, an evolutionary biologist then at the University of California at Berkeley, was at that time the first to propose that morning sickness could be an evolutionary adaptation that guided pregnant women away from pungent vegetables such as cabbage, cauliflower, and Brussels sprouts, which potentially contain toxins that could affect unborn babies. (Meat, which throughout human history has been especially likely to carry disease, is a particular trigger for aversions.) Profet's work was controversial at the time, but has steadily become less so, as scientists collecting data from all over the world have found some variation of morning sickness common to most pregnant women, beginning in the first trimester. Women who suffer nausea are actually less likely to miscarry, and those who vomit are even more protected.

The heightened sensitivity to food smells seems to vanish soon after the mother gives birth. Yet a mother's sense of smell may play an important role during her first encounter with her baby, especially if human moms, like mice, receive a boost of olfactory neurogenesis at delivery. The reason could have to do with what's called *imprinting.* An imprint is a feeling remembered for a lifetime: the first strong strand in the web of sensations binding mammal mothers to their children.

Imprinting is critical for ewes; in a crowded, jostling environment, these animals must learn to recognize their lambs before they lose them in the herd. Ewes also need to avoid letting eager interlopers suckle in their babies' place. Thanks to the sheep's sense of smell, most ewes recognize their own lambs within thirty minutes of birth. But when deprived of this sensory output, they'll let any lamb nurse at their teats. And they'll do the same if deprived of the smell of their own lambs immediately after birth.

Alison Fleming, the University of Toronto psychologist, compares this to the way some human mothers quickly learn to recognize their infants by other than visual clues, even though they're much less likely to lose them in a crowd. Within one week after birth, a mother can usually distinguish a T-shirt her own infant has worn from one worn by another baby. Even mothers with only limited exposure to their infants before they take a smell test can recognize their babies' clothing's odor by twenty to forty hours after delivery, supporting the idea that smell plays a strong role in bonding for humans, just as it does for animals.

Fleming isn't convinced that new mammal mothers smell any better than the rest, but believes that something happens in an ancient part of the brain, once a female has a baby, that makes infant odors more *attractive.* Two to three days after delivery, new human mothers are more likely to enjoy baby smells than nonmothers.

This effect may last only as long as the mother keeps close contact with her young. With rats, that's just a few weeks, before the pups are weaned. But human parents keep their children nearby for much longer, which may help explain why I still find the smell of my nine-year-old son Joey's head so intoxicating.

Audio Input

Motherhood may also enhance a woman's hearing in some way, though the case for this isn't nearly as powerful as the one for the sense of smell. No one at this writing, for instance, had found any evidence that new cells are generated in the auditory cortex, which is roughly akin to hearing as the olfactory bulbs are to the sense of smell. Each of these brain centers is responsible for how we perceive and interpret signals from the outside world.

What scientists do know is that the female brain starts out more sensitive to sound than the male's. And many mothers are convinced they quickly grow *more* sensitive to the sounds of their own babies, and more skilled in interpreting what they mean. You may know the feeling: You're at a party where a bunch of kids are playing in another room. You hear a series of cries and shrieks, but the sound that finally gets you out of your chair is your own child's voice, sufficiently changed to hint that things are getting out of hand. "Never before have I been able to hear the difference among babies crying," says Ulrika Engman Felder, a yoga teacher and recent first-time mother. "Now I know right away what age they are, and if it's my Anjali or someone else's baby."

Jeffrey Lorberbaum, the South Carolina brain-scanner, was impressed to discover that all the mothers and fathers he tested could identify their own newborn's cry compared to others. This intimate learning apparently starts early. One study found that almost 60 percent of mothers rooming together in the hospital with their infants and other mother-infant pairs reported waking up to their own infant cries on the first few nights after the birth, after which the percentage rose to 96 percent. Other research, performed about a decade before brain imaging was used, showed that patterns of a mother's heart rate change in response to her own baby's cry, but not to the cry of another mother's baby.

This seemingly "instinctive" response begins with some help from hormones, Alison Fleming believes. Glucocorticoids, stress hormones, are elevated in new mothers and fathers, making them generally more alert and ready to react. But over time, as this chemical influence subsides, the primary parent's intuition boils down to a matter of learning and expectations. Mothers, who, again, are customarily more regularly in closer contact with their children than fathers, are early on exposed to many more repetitive experiences with the child; through these experiences, the mothers learn how to differentiate between a shriek of joy and a yowl of pain. Mothers may also be driven to pay more attention by a greater sense of responsibility, for reasons of personal taste, cultural pressures, or mere disproportional investment (starting with lugging around a pregnant belly for nine months) making them generally more tuned-in to how their children are faring. "Think of it as if you're getting ready to buy a car," suggests Fleming. "Suddenly you notice cars all around you. The difference is your brain is more *activated* by auditory cues."

As with her sense of smell, a mom's pricked-up ears early on make her more perceptive and aware—smarter—about her own child. Under ordinary circumstances, all this perception and learning becomes part of her general feeling of extraordinary attachment to this new being. Later on, as I argue in subsequent chapters, that strong attachment itself can help make her smarter about the rest of the world, in part by keeping her brain enlivened as she copes with the variety of mental challenges embodied in a growing child.

Radar Vision

Of all the ways we perceive and interpret our environment, sight seems somehow the most "human"—that is, the least instinctive and the most rational. It's the show-me sense, after all. Yet, when it comes to mothering, you may be using your vision in subtly new ways, including the kind of ways that lead people to talk about mothers as having eyes in the back of their heads.

One change you may notice almost immediately is the way you look into your baby's eyes. The manner in which most adults, and most particularly parents, respond to an infant's gaze is dramatically different from normal behavior. It's a longer, deeper stare, very much like the stare of new lovers, which is as automatic a departure for most of us as the way we adjust the volume and pitch of our speech when we talk to babies.

If your infant seems also to catch your attention in uncanny ways, that's also a sign of a basic change in perception, and one we seem to share with other mammal mothers. In Virginia, Kelly Lambert says she has seen her pregnant lab rats become more tuned-in to visual stimuli, at least those stimuli resembling rat pups:

> One day when I was standing there watching them in their cages, I noticed that one was doing something odd. The late-pregnant rat had gently placed her tail in her mouth and then backed into the corner of the cage, where she released it. She then went on with her business, noticed her tail again, and did the same thing. It was bizarre, but I quickly realized that their tails are pink and hairless—just like the pups that would be arriving in the next two days. It seemed that her brain was becoming hypersensitive to pink, hairless stimuli, meaning she had to retrieve every such stimulus she saw and take it to her nest. I've spoken to other scientists since then, and they also have observed this behavior.

The long-cherished notion of Mom Radar is most important when it comes to sensing danger in the environment. When my children were babies, and I was still living in Rio, I occasionally took them to a playground next to our apartment. The lovely setting beneath the famous Corcovado mountain overlooked fishing boats on Guanabara Bay. But the grounds left a lot to be desired. Although they were safe enough by day, each morning I'd see evidence of the previous night's debauchery, such as used condoms and syringes scattered under the park benches. During my childless life, I might never have noticed them. Yet now I felt myself visually sweeping the grounds, just as a cop's practiced eyes sweep the inside of your car after he stops you on the freeway.

Obviously, the ability to quickly perceive and interpret threats has been crucial for human survival through the millennia. Just as our ancestors watched out for saber-toothed tigers, modern-day pedestrians must occasionally leap out of the path of oncoming Hummers. Yet, at this writing, just one piece of research seems to throw light on how a mother's visual ability may be enhanced once she has begun to take responsibility not only for herself but for her dependents. It came, oddly enough, from J. Galen Buckwalter, the Los Angeles psychologist who made national news with his study of pregnant women's purportedly impaired short-term memories. "Reporters missed one pretty interesting tidbit," Buckwalter says. Although his pregnant subjects seemed to have declined in their ability to repeat lists of words backwards, they performed "amazingly better" than their nonpregnant peers on tests of visual perceptual ability. This is additionally interesting because visual perceptual ability is normally a skill in which males, on average, excel. And it would seem to suggest that when you fly, you shouldn't worry if your pilot is pregnant, even if she forgets your name.

Buckwalter, who has a more sanguine view of the nexus of motherhood and mental resources than the reports on his study suggest, says he was unsurprised by this finding. "It makes no sense that evolution would strip women of essential cognitive reserves when they are so physically vulnerable at such a critical time for reproduction," he says.

What Touches Your Heart and Brain

In this sensory round-up, I've saved touch for last, though it may well be first in putting the magic in motherhood. A mother's hands, arms, and chest, as she cradles, carries, and caresses her young, are her most direct and loving means of communication, just as her suckling infant talks back with lips and hands, egging her on, before he knows a single word. Fleming has shown this with powerful evidence from the world of rats. In one poignant experiment, she separated new mothers from their litters immediately after delivery by placing a clear plastic perforated sheet between them. This allowed the moms and pups to see, hear, and smell each other, yet forced them literally to lose touch. As what otherwise would have been a pillar of the new relationship toppled, the rat moms' ardor waned. On tests ten days later, they responded to their babies much as virgin rats would: with fear and loathing.

In contrast, I can still vividly recall how my youngest son, Joshua—back in the days when his entire body was roughly the current size of his head—used to curl up on my chest at night to sleep. He snuggled over my heart, calmed by its beat and the warmth of my body. I lay on my back, sharing his warmth and breathing in his seductive baby scent. Somehow, both of us managed to sleep through the night, without one toss or turn. We awoke at the same moment, gazing into each others' eyes. As I write this, Joshua is almost six years old, but the sensation seems as fresh to me as if it had left a physical imprint.

As it turns out, it did—a state of affairs that requires some explanation, involving the story of a little man. It's not a real man, but a cartoon figure with distorted features, known as the *sensory homunculus,* an ingenious image that helps people studying the brain understand how it relates to the outside world. The drawing maps out what parts of the cerebral cortex, the thin, wrinkled layer made up mostly of specialized nerve cells ("gray matter"), covering the two hemispheres of the rest of the brain, are engaged when the body receives signals through the five senses. It thus has outsized lips, eyes, fingers, and tongue, representing the importance of these body parts in helping us take in information. In other words, the brain real estate accorded to each body part is relegated according to its role in helping us navigate the world.

Studies on animals strongly suggest that motherhood re-plots this map of the brain. When Joshua lay on my chest, he had a direct impact on my sensory homunculus. With repeated input from suckling and nestling, my chest, which used to be a purely aesthetic part of my personal repertoire, had acquired a leading role in my nurturing of another human being—and also in the way I was imagining myself, interpreting the world, and learning to behave.

In 1994, two neuroscientists, Judith Stern at Rutgers University and Michael Merzenich at the University of California at San Francisco, provided stunning evidence of this impact when they showed that in the cortex of a mother rat, the area devoted to the trunk, or chest, had actually doubled in size while that rat was nursing her young. Both Stern and Merzenich have little doubt that the same kind of thing happens in humans: "We were showing the response of the brain to experience," Stern says. "And this was the tip of the iceberg of possible brain changes for a mother."

Merzenich, who has taken a break from teaching to develop therapeutic programs commercially for brain-impaired children and elderly people, emphasized that what's happening is mostly a change in the way you interpret perceptions. As you touch your baby, and he touches you as he shifts in your arms, you're receiving subtle yet powerful information about his personality, emotions, and relationship with you. "You're getting more salient, specific feedback, as you're enlivening the surfaces involved in this engagement," Merzenich says.

Plastic, Elastic Brains

Merzenich and Stern's study of nursing rats claimed a place on one of history's most exciting frontiers of brain research, one inspired by a notion presciently expressed by George Eliot, in the mid-nineteenth century, when she said, "Our deeds determine us as much as we determine our deeds." It wasn't until the "Decade of the Brain," in the 1990s, that scientists established that experience indeed concretely helps reshape our cerebral equipment. Today, although much remains to be understood about this enticing phenomenon, it's widely recognized that the human brain, previously assumed to be set in stone by adulthood, can grow and change throughout a lifetime in response to new stimuli.

Whereas once it seemed we had to reconcile ourselves to the sad fate of a steady loss of brain resources as we grew older, we now know that new neurons and connections are formed all the time, the phenomenon known as *plasticity.* What stimulates us in a sense re-creates us, creating new and stronger pathways between synapses. These strengthened nerve-to-nerve connections are what learning is all about. Practice may not always make perfect, but it always makes *different.* And perhaps at no other stage in life are women as thoroughly re-created, in this way, as when they must adapt to caring for a new, dependent being. "There are certainly times in life when there are windows in brain development, when the brain is more plastic than at other times. Motherhood is one of those times," says David Lyons, a Stanford University primatologist.

When Stern and Merzenich reported their findings on the plastic brains of mother rats, the implications of plasticity for humans were only just beginning to seep into popular conversation. Yet the ideas behind them had been gathering force for nearly two centuries.

As early as 1819, the Italian anatomist Vincenzo Malacarne suggested that experience could change the structure of the brain. In 1874, Charles Darwin observed that the brains of domestic rabbits were smaller than those of rabbits living in the wild. He theorized that the difference might be due to the impact of living in confinement over several generations, deprived of all the stimulation of having to hunt and avoid predators. Then, early in the 1900s, the Spaniard Santiago Ramon y Cajal, who won the Nobel Prize in medicine, argued that "cerebral exercise" could make new and more connections between nerve cells.

Some six decades later, Marian Diamond and her fellow researchers at UC Berkeley made their important discoveries about how the brains of rats are enriched by stimulation. In the 1980s, another classic series of experiments, by Merzenich and colleagues, showed that an adult monkey's motor cortex could change in response to changing circumstances. When Merzenich severed a nerve conveying information from a part of a finger, the cortical patch that had previously responded to it adjusted to respond to other regions of the monkey's hand.

By the 1990s, excitement was mounting about the possibility that plasticity in humans could help the brain recuperate from illness and injury, or even help keep a brain working efficiently as it ages. Using proliferating new technologies, neuroscientists were just beginning to confirm that human brains were as malleable as those of other animals. In 1993, in one of the earliest findings of how repeated sensations can physically change the human brain, Alvaro Pascual-Leone, then at the National Institute of Neurological Disorders and Stroke, looked at the sensory cortex of fifteen skilled Braille readers. He found, as he'd expected, that the cortical area devoted to inputs from the finger most used for reading was much larger than that of sighted people's fingers.

Merzenich says you could expect to find similar differences if you looked at the brain of an experienced watchmaker and noted how much area was taken up by inputs from his eyes; or that of a flautist when looking at the area assigned to his lips. "We are all specialists," he says. "You're refining the machinery of the brain as you practice any skill."

As researchers have probed more deeply in recent years, they've discovered surprising plasticity in brain regions other than the sensory cortex. For instance, thanks to MRIs, we now know that people who take up juggling—a favorite metaphor for the way busy moms conduct their lives—increase their neurons, or gray matter, in areas of the brain known for storing visual information, over the time they're learning.

What visible changes might we expect to see, deep inside a new mother's skull? Kinsley and Lambert's research on rats and the Swiss experiment brain scans described in Chapter 3 offer some clues, but so far there's little data: The studies on humans that would answer this question are only just beginning. What we do know, as we embark on the powerfully emotional course of learning that is motherhood, is that within the crazy forests of wiring in our brains, some connections strengthen, with repeated practice, and others weaken as we use them less. (Think back to that analogy of walking through the woods the same way again and again, each time making a clearer trail.) Circuits that activate as we show empathy and affection, for instance, may become stronger and respond more readily. (For more on this, see Chapter 8). The neurotransmitters, or chemical messengers, coursing through our brains might be produced in greater quantities, carrying sensations of pleasure that reinforce new habits. Some areas—perhaps including the hippocampus, as we depend ever more on our memories—may be acquiring more gray matter.

Merzenich points out that training a brain in just one new task, equivalent in difficulty to using a spoon, directly results in changes involving hundreds of thousands, or even millions, of neurons, while altering the intensity of possibly hundreds of millions of synaptic connections. "A mom's infant-delivering and infant-carrying life is *full* of new experiences and learning," he says, producing "myriad and substantial change that can be expected to distort her poor brain, from that point forward." Moreover, motherhood is an especially powerful experience because it involves "learning under high stakes conditions, which is just the sort of learning that drives change in the brain," he says. "You are relating to another person like you never related to anyone before. In a sense, you are one person, and it's pretty educational to have another human being extending into the makeup of your own self."

In the case of a nursing mother rat with a remapped sensory cortex, Merzenich and colleague Judith Stern assumed the changes would reverse within a few weeks as the rat gradually lost physical contact with her young. "It's like the piano player who doesn't practice, and like you may not want to go to a brain surgeon after he has just come back from a long vacation," says Stern. At the same time, she and other experts agree, some kinds of learning are so emotionally powerful that they take much longer to fade. Think of a grandmother "instinctively" rocking a baby, or remembering how to swaddle a newborn, or being able to see past a child's, or another adult's, bluster. "It's perhaps less like playing the piano well than like getting on a bike. You never really forget," says Robert S. Bridges, a Tufts University neuroendocrinologist who has studied maternal behavior for more than three decades.

For human mothers, then, the sensory-rich, emotionally charged first weeks and months of life with a newborn potentially serve as the foundation for other relationships and experiences in years to come—the start of a virtuous, or, if things go wrong, vicious circle. "Every time I act or react, I receive associations, and I associate these with myself," says Merzenich. "I do that billions of times, and in the process, I'm creating myself as a context , based on all my experiences. . . . When I think, I associate myself as a thinker. When I act, I'm associating myself as an actor. In this way, we inherently are constructing ourselves, and our sense of self, and how it extends from us to others."

Accordingly, he believes, some changes women undergo as mothers would be "permanently self-sustaining, if the experience of motherhood results in changes in behavior that are heavily practiced thereafter. For example, if the attachment to my child cultivates my strong positive empathy for children and for life in general, I likely permanently exercise that on a level in which it is sustained through repetition to the end of life."

Merzenich's ideas are as controversial as they are exciting. Known for his scientific brilliance, he has also been called a "zealot" of plasticity. Still, in the early years of the new millennium, his theories were attracting some powerful support, from human as well as animal studies. Jeffrey Schwartz, at the University of California at Irvine, for instance, has reported how people who have undergone training to cope with obsessive compulsive disorder have changed their brains as they have changed their thinking. From its first heady sniffs and embraces, motherhood indeed can become a practice of wholly new ways of relating to other human beings. In later chapters, I'll explore what this means for the development of motivation and social skills, and in coping with stress. For now, let's look at how mothers can learn to be more resourceful and focused; in short, like Kinsley's cricket-catching rat mom: efficient.

CHAPTER
5

Efficiency
How Necessity Is the Mother of Multitasking

I have a brain and a uterus, and I use both.

PAT SCHROEDER,
THE FIRST FEMALE MEMBER OF
CONGRESS WITH YOUNG CHILDREN

EARLY ONE SUMMER morning, Kathy Mayer, M.D., the regional chief of internal medicine at Colorado Permanente, was pulled over by a traffic cop for making a left turn after the arrow had changed. "This ought to be interesting," Mayer said to herself, as the officer approached her car. Watching him in her side-view mirror, the blonde, baby-faced physician flipped off the switch to the gurgling black machine, the size of a briefcase, resting on the passenger seat. As was Mayer's habit in the hectic year after the birth of her twin girls six months earlier and her subsequent promotion to a job that had her driving all over Denver most days, she'd been using her hands-free electric breast pump, plugged into her car's lighter outlet. Sometimes while she drove and pumped, Mayer, thirty-four, would also return calls on her cell phone. On this day, however, she was merely driving and pumping, which was probably a good thing, because that alone was enough visibly to rattle the cop.

"He looked at me for a really long time," she relates. "My husband later said he probably thought I was strapped to a bomb." As the officer slowly made sense of the scene, he became visibly embarrassed, despite the discreet navy blue bib covering Mayer's décolletage. "He just said, 'license and registration,' then came back to the car and talked really fast, like: Here's-your-ticket, you-can-go-to-court-if-you-so-choose," Mayer said. "I thought it was hilarious."

Maybe so. But the fact that many U.S. jurisdictions have laws against driving while talking on a cell phone, but not against driving while pumping breast milk, may merely signify a legislative lag. Because, as most mothers know too well, there are only so many hours in a day.

Habits of a Multitasking Mind

While the Mommy Brain cliché suggests that women blunder through motherhood, we may be at our most efficient, and for good reason. As Craig Kinsley likes to point out, the Greek goddess Artemis was the patron of both childbirth and the hunt—a mother and a provider. In our modern era, in which one of two marriages will likely end in divorce, many mothers feel especially keen responsibility both to nurture and to support their children financially. Often, our mental resources seem to expand to meet this challenge, and we find ourselves becoming more expert in doing more than one thing at a time. When Jeffrey Lorberbaum and Samet Kose interviewed twenty-nine new mothers from six to eight weeks postpartum, asking them to rate themselves on how they might have changed, the women reported a highly significant increase in visual and auditory alertness, and in their ability to multitask.

To be sure, there's multitasking and there's *multitasking*. In 1994, Sheila Gilbride thought she was already pretty good at doing two or more things at once, but when she gave birth to premature triplets, she realized just how much there was to learn. "People would say all sorts of outrageous things to me, like 'If they were mine, I'd slit my throat,'" she recalls, laughing. Instead, Gilbride figured, "You have to do what you have to do."

The first thing she had to do was feed them, a goal complicated by the threesome's initial need to eat every three hours. "I was home alone—

my husband went back to work—and I knew I had to feed them all at once, or I wouldn't get any sleep at all," Gilbride recalls, with painful clarity, ten years later. And so she would prepare three bottles: "I'd prop up Brianna, who could sit alone, on the sofa, within reach in case she choked. I'd have my legs up on the cocktail table and put Kayla on my lap with her head on my right knee and my right arm around her. Ryan, the boy, was having trouble with projectile vomiting, so he'd be wrapped in a towel and leaning up against my left knee, with my left arm around him."

When we spoke, Gilbride was working full-time as a senior executive for the pharmaceutical company Pfizer in New Jersey; she had a fourth child to care for, and her evenings were partly taken up by studies for her MBA. But by then she was practiced in time management: "What's remarkable is the fact that you get through it and learn from it," she says. "You get your priorities straight because you have to. This experience made me a lot more organized." Everything is relative; Gilbride recently met a mother of quadruplets, and couldn't stop herself from repeating the classic refrain: "I don't know how you do it." The mother responded, "Ah, you moms with triplets, all you do is complain," causing Gilbride to reflect, "It's true. I look at moms with twins and I think, no big deal!"

Popular wisdom, grounded in evolutionary theory, dictates that women, and particularly mothers, excel when it comes to splitting their focus. From the Pleistocene's hunting-and-gathering fields to today's outdoor play structures, a mother's capacity to combine tasks has been key to her children's survival. "You watch two women at a playground, and they can be having a very interesting conversation, while they're also being very protective of their children," says John Morrison, a neurobiologist and father of two, who works at the Mount Sinai School of Medicine. "The fathers either have to use all their attention to follow the kid around, or the kid falls off the jungle gym."

Still, when we talk about mothers multitasking, we often mean something grander than, say, a woman who is cooking lasagna while talking on the phone. More than a mere juggling of tasks, it's a way of life entailing an ability to focus on the essential, to ignore the irrelevant, and to accomplish a lot more in a given time. In a word, it's efficiency. "You prioritize better because you *have* to: It's pure survival," says Sandy Wells, an elementary school teacher and the mother of two young boys.

Children, like fear, have great power to focus the mind. You become trained in triage from the first few days of their lives, responding to the urgency of feeding, burping, protecting from sudden falls, and in so many other ways, as well, keeping another being alive. If you're lucky, the stress that keeps you on your toes stays at an optimal level, as described by something called the Yerkes-Dodson law. Its inverse-U shaped pattern relates arousal, which we often think of as mental stress, to mental performance. Some stress is vital: You won't learn if you don't care. But too much stress means learning becomes a low priority. Halfway to excess, you'll be at your peak.

Mothers who grow used to operating at that peak may be training and sharpening their capacity to pay attention, a mental skill that the Harvard psychiatrist John Ratey, an expert in attention and the lack of it, defines as "much more than simply taking note of incoming stimuli. It involves a number of distinct processes, from filtering out perceptions, to balancing multiple perceptions, to attaching emotional significance to them." This useful habit can be reinforced through the release of the neurotransmitter dopamine, which carries a sensation of reward through the brain and is released during mildly stressful situations (and during pleasurable activities, such as sex and good dining), says Ratey. He calls dopamine "the attention, learning, and motivation neurotransmitter," and says it may be the link between pleasure and long-term memory. "An increased level of dopamine improves your attention system and uses many more resources of the executive function than you had earlier," Ratey says. "If you build up these resources, you can have them available afterwards."

Ratey says it's reasonable to expect that new mothers might sharpen their attention through this and other means: "When a mother gives birth, you want her to be really smart, as smart as she can be. You want her to know about the territory around her and remember things about her kids, and be in a prime state to function. It has to lead to an ability to pay more attention to the outside world."

Being in the Moment with an Eye on the Clock

All this is not to deny that children on occasion can destroy your concentration like little else, especially when you're on the phone. Jane Lubchenco, an internationally distinguished zoologist at Oregon State University, recalls that when her two sons were young, she often found herself recalling the Kurt Vonnegut short story "Harrison Bergeron," a tale about an invented future, when equality among humans was enforced by a Handicapper General who would make sure that no one took advantage of natural strengths. Gifted ballerinas had to wear heavy weights, and people with above-average intellects had to wear a mental handicap radio in their ears that sent out a sharp noise every twenty seconds or so. "My husband and I would joke about how that's what our kids did," she says.

On balance, however, Lubchenco maintains that her determination, together with that of her husband, to be present and spontaneous with their children ultimately made her smarter—in terms of being more present, focused, and organized. "We tried consciously when we were with our kids not to be thinking about everything that wasn't getting done but to treasure the time with them . . . it was being in the moment as opposed to constantly being frustrated." This practice carried over to work, she adds. Accepting the limits on her time, she became more deliberate about how she spent it. She chose to write fewer but weightier papers instead of numerous lesser ones. "I'd put the less substantial ones in a 'maybe-someday bin' because I figured I would never be able to keep up with all of them," she says, "and that conscious decision in the end was a good career choice."

The sensation of limited time is a fundamental aspect of motherhood. Your children's need for you is so enormous that you may feel there are never enough hours to satisfy it. And you may also be struck by how quickly they seem to grow—unavoidable evidence of the speed of passing time. The British novelist Rachel Cusk writes about what it felt like to "purchase back" hours from a babysitter, only to find that "they were hours whose crazy ticking could be heard. Living those hours was like living in a taxi cab. Working in them was hard enough; pleasure, or at least rest, was unthinkable."

Although your level of anguish may be higher, so, often, is your capacity to hunker down. Working mothers often become masters of time management, trained by those implacable deadlines. "Your whole day is shaped by the fact that you know you can't hang around in the lab into the evening," says Tracey Shors, the Rutgers University behavioral neuroscientist and the mother of a four-year-old son. "The daycare is closing, and you gotta go."

The famed author of the *Harry Potter* series, J. K. Rowling, has described how she combined writing her first *Potter* book with skirting poverty and single-handedly caring for her baby daughter. One secret was making full use of nap times. "I used to put her into the pushchair and walk her around Edinburgh, wait until she nodded off, and then hurry to a cafe and write as fast as I could," Rowling told a reporter. "It's amazing how much you can get done when you know you have very limited time. I've probably never been as productive since, if you judge by words per hour."

Many new mothers quickly start to notice all the errands and phone calls that don't have to be dealt with right away, or ever. "I can't believe the stupid things I was obsessed with before I was a mom," says Mayer, the Colorado Permanente doctor, who fits parenting in between working forty hours a week away from home and ten to twenty hours more in her home office.

Racing through their days, mothers often report feeling a change in their basic thinking processes, as if their brains were working on several tracks at once. The fictional Kate Reddy, in Alice Pearson's novel *I Don't Know How She Does It,* demonstrates this with her urgent to-do lists for work and home: "Teletubbies cake where? Pelvic floor *squeeze.* Return *Snow White* video to library! Emily school applications get organized. Be nicer, more patient person with Emily so doesn't grow up to be needy psychopath. Quote for new stair carpet . . . "

Says the Virginia neuroscientist Kelly Lambert:

I have more cognitive programs running in the background. All of a sudden, for no apparent reason, I remember things like "Did I turn in Skylar's permission slip for her field trip?" or, "We need more snacks for gymnastics breaks," or "We need to invite so-and-so over because they had us over two weeks ago. These reminders just keep popping up in the foreground, and the weird thing is that they're not prompted by obvious stimuli, but they appear at the appropriate times to remind me to do them—like there's some grand office manager sitting in there sending me the mind's equivalent of those [Microsoft] Outlook reminders that pop up on my screen when I have a meeting—except I don't hear that "ding" sound, yet.

What a Difference a Gender Makes

As research on gender differences has taken off since the 1980s, some experts have come to believe the female brain may be set up from birth for extra efficiency in accomplishing multiple tasks. It's important to point out that most notions about male and female brain differences, and what they mean, remain steeped in controversy. What's more, the Big Message of the 1990s—that changing the way we think and behave can make concrete changes in our brains—complicates this discussion. Just as with rearing children, nature *and* nurture combine in brain development.

That said, there are a few intriguing physical differences between men and women's cerebral equipment that some scientists are convinced lead to differences in the way the two genders think. The most noticeable difference is overall size: The female brain, on average, is about 15 percent smaller than the average male brain. This is a particularly incendiary topic; a few modern scientists have suggested that brain size is related to intelligence, implying that women, in general, are dumber. Other experts have challenged the larger-is-better view by pointing out that women's brains are more tightly packed with neurons, *ergo*, more efficient machines.

Other types of comparison yield more minute physical differences, but often just as much speculation. The neuroanatomist Marian Diamond notes that the female cerebral cortex is, on average, slightly more symmetrical in thickness than that of most males. She suggests this difference evolved to optimize a female's ability to rear and protect offspring, roles that "challenge her to go in many directions"—taking in information from multiple sources, often at the same time. In contrast, she says, the male's slightly more asymmetrical cortex may have evolved through the prehistoric division of labor in which his jobs—including finding food and defending territory—were more singularly focused.

Although much remains to be understood about this potential relationship between structure and function, recent *f*MRI scans have established that women do seem to use more of their brains at once, compared to men, in response to some stimuli. In one study, researchers looked at the brain activity in ten men and ten women who were listening to an audiotape of a John Grisham novel. In all cases, the men's brains showed activity exclusively in the left temporal lobe, but women's brains were active in both the right and the left temporal lobes.

One other physical difference that is often cited in comparisons of male and female brains has to do with a highway of nerve fibers connecting the left and right hemispheres. Called the *corpus callosum,* the "calloused body," it's an arched structure running midline from the back to the front of the head. Some studies have found that a female human's corpus callosum is on average slightly thicker at one end than a male's, in proportion to total brain size. (Another, smaller connecting bridge is called the *anterior commissure,* found to be on average about 12 percent larger in women.) Size does now seem to matter, as many experts suspect the woman's more generously equipped brain is more efficient in relaying information back and forth between the intuitive right hemisphere and the businesslike left. Helen Fisher, a Rutgers University anthropologist and the author of *The First Sex: The Natural Talents of Women and How They Are Changing the World,* writes that "women's well-connected brains facilitate their ability to gather, integrate and analyze more diverse kinds of information," an aspect of what she calls "web thinking." Women, she says, "gather more data from their environment and construct more intricate relationships between the information. By contrast, men tend to compartmentalize—to get rid of ancillary data and focus only on what they regard as important."

Mark George, the Medical University of Southern Carolina brain-scanner, suspects that, in practical terms, the difference may help account for how a guy can stay glued to the television through the ninth inning while his progeny are loudly murdering each other in the next room. He feels more certain that the different way that women use their brains gives us a valuable resiliency. "It has been known for almost a hundred years that if you have a stroke that damages a certain part of the brain, men tend to have more functional disability than women," George says. "You match ten men with ten women, all of whom have had a stroke, and the men would have greater deficits, such as not being able to raise an arm or leg, but the women would not, which seems to imply that women have a broader distribution of function. Things aren't so pigeonholed."

Off to the Races

Back in Virginia, Kelly Lambert and Craig Kinsley suggest that a female's innate mental capacities switch into high gear when she becomes a mother. They base this in part on their ground-breaking research showing that rats and monkeys, once they reproduce, improve basic memory and learning skills—core strengths for multitaskers of all species.

In the rat studies, the two researchers compared the performance of "multiparous" and "nulliparous" females—those who had delivered, nursed, and weaned two litters versus age-matched rats who had never mated—as they looked for Froot Loops, a rat delicacy, hidden in an eight-armed maze. The object of the test was to remember the location of the treat, and as it turned out, the mother rats learned faster, making many more correct choices in the first six days, as both sets of rats gradually mastered the test.

Since then, one of Lambert's undergraduate students, Anne Garrett, has found something similar in a small study of marmoset monkey moms tested for their skill in finding Froot Loops hidden on a board hung with small rubber change-purses of various colors. Green purses contained two Froot Loops; blue purses, one; reds, none. "It's a lot like shopping," says Lambert, "and the moms really seem to know what they are doing."

Kinsley believes evolutionary pressure, favoring the survival of the fittest, helps explain the speedier mom-rat learning curve. While nursing hungry litters, rodents must be prepared to roam far from their nests—navigating unfamiliar space—and remember how to find their way back to their hungry pups in time to feed them. "Imagine that you have a mother who has just given birth to ten or twelve pups," he says. "Unless she has been fortunate to be situated right in the middle of food and water, anything that brings her back to her nest more quickly would be expected to be preserved or enhanced because it gives her the advantage over a female who hasn't had pups." Limited as they are to rats and marmosets, Kinsley and Lambert's findings plainly offer an optimistically contrarian message for humans. Parenting, they believe, boosts a mother's brain—and not just when it applies to direct care for her babies, but in ways that help her cope more efficiently with the outside world.

As the media in the United States, Europe, and Japan repeated these findings—and late-night talk-show host Jay Leno joked that "only after a mom gets pregnant does she realize what a jerk she has married"—many mothers felt a pang of recognition. The rodents' improved foraging skills seemed to be a primitive version of the way modern parents run about, retrieving diapers at the drug store or data and a paycheck in the course of their information-economy jobs. Lambert, following the *Nature* debut, even designed one rat trial specifically to test motherly multitasking. In this test, rats who had never encountered each other before or been involved in competitive tasks were placed in a round maze the size of the bottom of a paint bucket. After initially locating where a Froot Loop had been placed, they were returned to their starting positions. They then not only had to remember the location of the reward but also had to respond to so-called social cues, in essence sizing up the competition. As it turns out, the multiparous mothers found the Froot Loop 60 percent of the time; the mothers with just one litter, 30 percent of the time; and the pupless rats, just 7 percent of the time.

The Virginia researchers believe that the improvements they've found in their mother rats' memories arise from a two-part process: the preparation of pregnancy, followed by the stimulating presence of pups. Remember that in pregnancy, a female's brain is awash with mind-altering hormones, principally elevated levels of estrogen. As described in Chapter 2, estrogen in some situations has been shown to boost brainpower. And researchers have long known that estrogen is most likely responsible for the forests of dendritic spines that Kinsley and Lambert found in their pregnant rat brains. Catherine Woolley, a neurobiologist at Northwestern University who also looked at the hippocampi of rats, found that dendritic spines increased significantly during the high-estrogen phase of a rat's menstrual cycle, the time, perhaps not incidentally, when the rat is most fertile and presumably needs some smarts to choose a mate. And John Morrison, the neuroscientist mentioned earlier in this chapter, has found that in monkeys, estrogen increases synapses in the part of the cortex that deals with attention and complex tasks.

Morrison believes that estrogen is important throughout a woman's child-bearing years precisely to make her sharper. "If estrogen can help females in not only their success rate at having offspring, but at raising those offspring successfully, then it's going to have enormous evolutionary pressure," he speculates. Might the high levels of estrogen in a pregnant woman's brain thus have some long-term, beneficial effects? Morrison suspects it's possible, and Kinsley believes it's likely, which is why, instead of looking at a pregnant woman as a slow-moving, slow-thinking "fetal incubator," Kinsley envisions a woman with a brain "like a hugely-powered dragster . . . wheels churning, smoke pouring from its exhaust, waiting for that green light to loose its horsepower, and go screaming down that track."

Brain Aerobics

The green light, in this vivid scenario, flashes with the onrush of smells, sounds, sensation, and overall panic embodied in an infant. (Anne Lamott once compared her newborn to a radio alarm clock going "erratically every few hours, tuned always to heavy metal.") All this stimulation is a potent force for brain plasticity, and, if you're lucky, it's also a source of enrichment because new challenges give your brain a healthy workout.

When Kinsley and Lambert examined the brains of their late-pregnant mothers, they found that their normal rate of cell replacement had slowed down, an event that could help explain why pregnant human females experience some brain shrinkage. But this process resumed and increased immediately after pregnancy, while the rats were nursing. Compared to pupless rats, the nursing rats also were found to have more glial cells, which support the neurons by importing energy and exporting waste products, and may also play a role in information processing.

Was simple experience making the difference? The Virginia researchers found support for this view when they compared the memory skills of three groups of rats: mothers, virgins, and foster parents. (Fosters are virgin females who have spent sufficient time with pups to lose their fear of them, and even start behaving maternally by licking, grooming, and retrieving the infants.) The point of this test was to find Froot Loops hidden in one of eight wells in a circular maze, and then remember to return to that same well, which would be restocked, in subsequent trials. Once again, the biological moms did best on this task. But the foster moms came in a close second. Even without the heavy dose of hormones, mere interaction with pups appeared to have made the fosters more efficient. The finding fit right in with the idea of enrichment: that stimulation makes you sharper.

Time for a quick reality check. It is certainly worth noting that, unlike a twenty-first-century human, a rat in a cage has few choices for stimulation. It can't go to the opera or study Japanese. The difference between being a rat in a cage alone and a rat in a cage with pups is quite significant, as is the consequent change in the brain. That said, the human brain, just like the rat's, might be thought of as a group of muscles: Use them or lose them. How and which ones we use help define who we are. Most parents are virtually forced to use certain cerebral talents, constantly and repetitively, as our children grow. "It's a lot like learning a foreign language," says Michael Merzenich, the brain development specialist. "When you're learning a foreign language, you're also learning about a whole culture, and music and literature. Here you are learning, close-up, about the human race."

Added to this is that, unlike the mental challenges of studying or working at a job, it's much harder to check out mentally for any length of time when you're on the spot for nourishing, nurturing, defending, discerning, interpreting, encouraging, and disciplining, all the while also trying to figure out how to make time for it all. You are usually *bound* to try to meet your child's needs, in ways that grow more complex over time. (How the heck do you swaddle a newborn, again? Is he going to fall now, or should you let him climb higher? Is there a God? Quick! Your five-year-old wants to know. And what to say to the camp counselor reporting that your ten-year-old just taught the kindergarten group to spell the f-word?)

Children continually make you flex your brains by way of a continuing series of progressive tests accompanied by the knowledge that, should you slack off, unlike taking a break from opera or Japanese, you risk living with the consequences far into the future. You may not be "good with your hands" when it comes to assembling a safety rail for a crib, for instance, but you learn. With lower stakes involved, I've found myself lately getting up to speed on laser tag, the care of pet rats, and how to prepare for a bar mitzvah, though so far I've drawn the line at "Yu-gi-oh!"

The task of explaining things to a child can also help adults to think more creatively. When James Dillon, the Georgia psychologist, surveyed parents and teachers about what they'd gained from contact with children, 11 percent of the parents and 47 percent of the teachers offered stories that Dillon categorized as relating to "cognitive flexibility," or being able to think about things in new ways, which interviewees themselves described as a process of becoming "smarter." Children generally think in concrete images, contrary to an adult's more detached, abstract style. Teaching them can make grownups pay more attention to this difference, and experiment with a more direct style. One teacher told Dillon that in teaching multiplication, she brought in a bunch of marbles to show her students how multiplying three by three was really like adding three sets of three marbles. "And you know, before then, I never knew what multiplication was . . . never really knew what that meant," she said. "It's now in my bones rather than just in my memory."

Often what this means is that you begin to think in a more spontaneous way, a process that starts for many mothers when they're obliged to figure out how to make their babies stop crying. "With a baby, much of your time will be spent in spontaneous activities, requiring that you reach blindly into your bag of intuitions and come up with a suitable reaction on the spot," writes Daniel Stern, a psychology professor at the University of Geneva, in his book *The Birth of a Mother.* "Some women adapt easily to this way of life, but for others it's quite difficult to operate in a realm where the rules are always shifting, and where you are not sure of the game to begin with. Even if you experience some difficulties, though, spontaneous reactions will become part of your new identity."

Kids present their parents with constant brain teasers, often together with the grim understanding that you will be endlessly annoyed for not coming up with the right answer. On a recent trip with my sons, we were driving down from a snowy mountain hill, listening to a Lemony Snicket CD, when Joshua, aged five, came up with an odd complaint.

"My voice is quiet!" he whined.

"I don't hear anything different," I tried to soothe him.

"No!" he yelled. *"My voice is too quiet!"*

"Joshua, it is *plenty* loud!" I responded testily.

He wouldn't stop complaining. Consciously, I just wanted him to quiet down so that I could look out the window and listen to what the villainous Count Olaf had in mind next for the Baudelaire orphans. But on another track, I must have been thinking, because suddenly I said: "Try yawning!" Instants later, he settled down, smiling, control regained. Being five years old, he'd simply been unable to distinguish a voice problem from a blocked-ears-due-to-high-altitude problem. That ended up being my job.

Relieved as I was to have come through with the answer, I wondered: Isn't all this testing likely, in the end, to leave you more depleted than enriched? And if enriched, might it not be solely in the specifics you know about your children, just as taxi drivers remember all the tiny streets and shortcuts? Or might there be a way you could be sharpening equipment you can use outside the home? For answers to these questions, I turned to Fred Gage, a neuroscientist at the Salk Institute for Biological Studies in La Jolla.

In 1998, Gage made neuroscience history by disproving the conventional wisdom that the neurons you're born with are all the neurons you'll ever have. He proved, first with rats and then with humans, that neurons are constantly being born in key brain centers responsible for memory and learning, their eventual usefulness depending on the quality and level of stimulation you get from your environment. When I told him of my quandry, Gage counted off four factors that can increase the chance that any given learning experience will improve the brain. It helps if the situation is *novel,* something you haven't encountered before. You should also be actively *engaged.* (You don't get new neurons by sitting around watching television, even if it's Jacob Bronowski.) It's important that the situation is *affective,* meaning you're emotionally involved. And it also makes a difference if the stimulation is *complex,* that is, challenging.

Novel, engaging, affective, and complex. As much as I've found all the *housework* involved in parenting to be routine, boring, draining, and dull, I know that when I stop to spend time with my children, this magic waits for me. And the more I seek and find it, the more I want to seek it again. Along with other scientists, Gage believes that once your brain changes in response to a given stimulus, it responds in a different way to the next challenge. This, for mothers or anyone else, could result in either a vicious or virtuous cycle, depending on your particular challenges and how you confront them. "What we do as individuals is feeding back to our brain," says Gage. "We have a lot more control over who we are than we often think."

Gray Power

Kinsley and Lambert tested mother rats well into rodent senescence, and they discovered something remarkable. The mom rats' gains in learning and memory, as demonstrated in the maze, were lasting up to twenty-four months, a full eighteen months past their last litter, and the equivalent of about eighty years of age for a human. The rats who had delivered more than one litter were still leading the pack. As Kinsley puts it: "Growing older gracefully may be a function of having had an active procreative life with lots of exposure to estrogen and offspring stimulation."

If you believe in a master design for nature, authored by God or by evolution, Kinsley's scenario makes sense, especially combined with discoveries by other researchers studying smart human grannies. The basic idea, which has been called the "grandmother hypothesis," is that grandmothers live long lives and need to be energetic and smart long after menopause, precisely to help their children's children survive. The poster child for this theory is any grandmother among the Hadza, a group of fewer than 1,000 northern Tanzanian nomads, who live by hunting and gathering, much as our ancestors did some million years ago. Kristen Hawkes, an anthropologist at the University of Utah, has found that the older female Hadzas provide noticeably more dinner than most of the rest of the group, leading her to believe that "the Grandmother Hypothesis gives us a whole new way of understanding why modern humans suddenly were able to go everywhere and do everything," as Hawkes has told the author Natalie Angier. "It may explain why we took over the planet."

So now we know there is something in reproduction that seems to preserve a female's smarts into old age, at least in rats. And we know that in at least one group of humans, old females have demonstrably good reason to be smart. One other thing that seems clear with many modern grandmas is that they rely on their grandchildren to help keep them smart by keeping them tied to the outer world. Unless you're very careful, growing older can make you love routine too much. "Think of someone who is eighty who gets a chance to move to another part of town," says Michael Merzenich, the brain plasticity expert. "It would be absolutely great for that person's brain to move, and take on all the challenges involved in that. But normally, of course, they would resist it."

Once again, however, a grandparent's love might prove stronger than such resistance to change, providing the mental force required to tackle, say, instant messaging. As *Maternal Thinking* author Sara Ruddick, now a grandmother herself, writes, "Grandparents in my social circles, now in their late fifties or older, have known illness, loss and death. In this context, grandparents need and want experiences that allow them to remain curious, appreciative and attached to the world that will survive them."

For grandmothers and mothers alike, children represent a constant pull toward that outside world—a world filled with intellectual challenges and with social obligations. As researchers have been finding, this social world is a particularly good world for your brain, potentially moderating stress and encouraging resiliency. This is the topic of Chapter 6.

CHAPTER
6

Resiliency

Reducing Stress, Enhancing Smarts

As pursed lips clamp tightly onto her nipples and tug, the little head gives a jerk, like a fish on a line whose movement secures the hook. But in this instant, just who is it that is being caught?

SARAH BLAFFER HRDY,
MOTHER NATURE

IF YOU'VE EVER breastfed a baby—and you weren't simultaneously wrestling with your nursing bra, talking on the phone, or yelping in pain at the emergence of teeth—you'll know there are few more intimate pursuits. Your heartbeat slows; your body temperature rises, and time seems to dawdle. You stroke his cheek; he gurgles and sighs; you lose yourself in his rapt gaze.

This kind of communion is unique. Yet for many new mothers, it's also a potent introduction to a new way of being while they're "hooked" into strong relationships including and surrounding their child. Motherhood, as the writer Erica Jong says, "made me join the human race."

In my own single thirties, I lived alone as a reporter in Mexico City. On one not extraordinary day, I remained in my home office from dawn to dusk, rewriting a magazine story. My only human contact was hearing Eloisa, the fierce Nicaraguan mother of three who lived in my house when I was traveling, and helped me keep it clean, repeatedly pass by my door. Each time she did, Eloisa, who couldn't fathom why anybody would choose to be so reclusive, made more obviously intentional noise. Toward evening, I heard her loudly muttering, *"Es terrible!"*

Maybe it was *terrible,* though I still sometimes pine for that solitude. The point is, it was *possible,* whereas, today, it's not. As the mother of young boys, my day is often noisily interrupted by their presence, their friends' presence, their pet rats' presence, and the knowledge that I must soon, for their sake if not mine, make contact with a virtual village of people from whom I'm seeking or paying favors, including teachers, room mothers, other mothers, pediatricians, swim team coaches, and pet store rat savants. I take great consolation in studies suggesting this might help make me smarter.

Sociability, as I shall detail in Chapter 8, can be a kind of smarts in itself, involving skills largely underappreciated as such until recently. But here I'm addressing more traditional aspects of intelligence, such as memory and focus. The famous Nurses' Study, involving more than 100,000 U.S. women and funded by the National Institutes of Health, has found that belonging to a strong social network correlates with better mental functioning. And a separate University of Michigan study of more than 3,000 adults demonstrates that people who maintain higher levels of social engagement, from spending time with others to talking on the phone, have better working memories and other cognitive skills.

Some of this surely has to do with the simple brain stimulation that you can get from interacting with others. Much also points to how spending time with other people can—at best—reduce your stress. We've known for years that stress reduction helps physical health, lowering the risk of heart disease and diabetes, and even strengthening the immune system. What's new is mounting evidence that stress relief can also help your brain, in the short and the long term. What's even newer is that mothers—who, by the old Mommy Brain model, would appear to be nothing but *victims* of stress—appear to have some advantages in moderating the stress we face. To understand these better, you need to know about a peptide hormone fundamental to mothering called *oxytocin.* C. Sue Carter, the behavioral neuroscientist and a leading authority on the topic, whom I introduced in Chapter 3, says the hormone's main importance is that "it hooks us into the social world." But there is also some remarkable recent evidence that oxytocin may directly help memory and learning, through potentially long-lasting changes in the brain.

The "Cuddle Hormone"

Oxytocin was discovered in 1906 by the English researcher Sir Henry Dale, who realized that it could speed up labor by stimulating uterine contractions. He named it using the Greek words *okus,* for "swift," and *tokos,* for "birth." A century later, scientists are just beginning to appreciate the hormone's range of other roles, amid excited hopes that it might one day be used as an antidepressant, a weapon against Alzheimer's disease, or even—according to one particularly optimistic researcher—a key to combating pollution and world poverty.

Oxytocin acts in two ways. Produced by the hypothalamus, it's released by the pituitary gland into the bloodstream, where it not only stimulates labor but prompts the "letdown" of milk once a baby starts to nurse. Neurons in the hypothalamus also lead back into the brain, where oxytocin acts as a neurotransmitter that carries information across synapses and sways emotions and behavior, promoting calm and, according to some researchers, cementing social bonds.

Male prairie voles given a shot of the stuff are converted from roving-eyed bachelors into nurturing husbands and fathers. Mice bred so that they can't be influenced by oxytocin fail to recognize other mice. Rats respond to doses of oxytocin by becoming more gregarious, less anxious, and more curious. In humans, oxytocin levels peak during sexual orgasm.

But surely oxytocin's most vital role comes in the realm of mother-child attachment. For new mothers and babies coming face-to-face for the first time, the hormone, which helps initiate maternal behavior, is "the endocrinological equivalent of candlelight, soft music and a glass of wine," as Sarah Hrdy has written. In rats, virgin females injected with the "cuddle hormone" abandon fear and hostility in order to lick, groom, and protect any pups in the vicinity. In humans, oxytocin in combination with endogenous opioid peptides—substances believed to cause the "high" experienced by long-distance runners and which are also released during mother-baby interactions—may help explain why an otherwise sensible adult can sit around for hours, saying such things as "Whozzat little pootchkin?"—and accordingly, why generations of new pootchkins survive.

Your sensitivity to oxytocin's power is one of the most fundamental ways you change as a new mother. In pregnant rats, new receptors for the hormone appear in the uterus, mammary tissue, and brain. In humans, such receptors have been found in the uterus and mammary tissue, making it reasonable to assume they might also be present in the brain, though for obvious reasons this has been hard to measure.

At Sweden's Karolinska Institute, Kerstin Uvnas-Moberg, a world authority on oxytocin, performed experiments in the early 1990s showing that breastfeeding women tend to be less reactive to stress hormones. Women from three to six months after delivery who have breastfed for at least eight weeks report they are less anxious, less physically tense, less suspicious, and less bored. They are also calmer and more sociable when tested for these traits than mothers of comparable ages who are not nursing their babies. Furthermore, Uvnas-Moberg has demonstrated that the degree of these new mothers' mellowness can be predicted by the level of oxytocin in their bloodstreams, which may correlate with the level in their brains.

Other scientists have found similar signs of a subdued stress response in nursing women. One research team, led by Margaret Altemus, a psychiatry professor at Cornell University, stressed ten lactating and ten nonlactating women by requiring them to exercise on a treadmill. They found the lactating women released only half the amount of three stress hormones, compared to those who were not nursing. Many, if not most, scientists assume these differences subside once the mothers stop nursing. But Uvnas-Moberg speculates that permanent change has occurred. She bases this in part on research showing that humans and other mammals respond more readily, physically, and emotionally, to their second baby than to their first. This is more a matter of having ridden that roller coaster before and knowing what to expect: The difference appears to be hardwired, even influencing the mothers' milk flow, which is generally freer the second time around. "With the first baby, something happens that changes you forever," Uvnas-Moberg says. Indeed, French researchers have found that the neurons that produce oxytocin in the brain of a rat are actually restructured by the act of giving birth and nursing. Other clinical scientists have shown that rats exposed to extra oxytocin as infants may have permanently reduced reactivity to stress, as well as lower blood pressure. These are just some of the reasons that Uvnas-Moberg believes a woman's brain is permanently changed through childbirth and breastfeeding, possibly making her nervous system less reactive. "Your antistress systems are activated," she says. "Definitely you are buffered."

The Mommy Edge

In the first years of the new millennium, two lab studies in particular offered tantalizing evidence supporting the idea that a mother who must learn so much, so fast, might actually have an enhanced brain to help her along.

One of these studies came from Okayama University in Japan, where in 2003 a team of researchers led by Kazuhito Tomizawa (and including three mothers) injected oxytocin into the brains of mice who had never been pregnant. Then, following Kinsley and Lambert's lead, they tested a group of these oxytocin-enhanced mice with a spatial learning task in which the rodents had to look for food in an eight-armed radial maze, with only four arms containing rewards. The oxytocin-boosted mice were much better at remembering which arms contained the goodies. The researchers then injected an antagonist to oxytocin into the brains of mice who had carried several litters. The rodents' performance on the maze task declined.

Tomizawa and his colleagues published their findings in *Nature* in April 2003. They wrote that they had found something called long-lasting, long-term potentiation, known as L-LTP, in the hippocampi of mice who had received the oxytocin injections, as well as mice who had given birth to more than one litter. L-LTP is a physiological marker of long-term memory formation, involving an actual increase in the efficiency of synapses that process information. So this was big news. Tomizawa enthusiastically speculated that oxytocin might one day be a therapy for age-related memory loss, such as that which occurs in Alzheimer's disease. "Readers of either sex now have an additional reason . . . to release some of their precious bodily oxytocin," wrote three scholarly reviewers.

That same year, Rutgers behavioral neuroscientist Tracey Shors presented her own findings showing that new mothers appear to be protected from stress that would otherwise harm their memories—a phenomenon Shors suspects also has to do with oxytocin. In previous work, Shors had shown that virgin female rats learn less efficiently after stressful experiences. She was also aware of the studies showing that breastfeeding women are less reactive to stress. So Shors designed a test with mother rats, whom she first stressed by restraining them inside a clear, Plexiglas tube; then, using a classic Pavlovian technique, she stimulated their eyelids with a tiny electrical impulse just as they heard a tone. When Shors measured how long it took the rats to learn to blink their eyes at the sound of the tone, she found the mothers learned more quickly than equally stressed nonmothers.

Shors believes this "unique response system" helps prevent new mothers from being overwhelmed, thus allowing them to take better care of their offspring. Her hunch about the "cuddle hormone" is based on the fact that oxytocin inhibits the release of the stress hormones called glucocorticoids. Glucocorticoids are one reason that sustained stress, over time, can damage the hippocampus. In particular, they weaken the capacity of cells to survive trauma from strokes and seizures.

The Stanford biologist Robert Sapolsky, an international expert on stress who has done groundbreaking work on glucocorticoids, agrees it's possible that oxytocin might be Mother Nature's way of lending a hand to mammal mothers. He illustrates this with a story about migrating birds. Birds that regularly fly from Baja to the Arctic can sit in freezing temperatures without any elevation in stress hormone levels, he points out, adding, "It's astonishing. And yet when you think about it, you realize that this is a normal day's work for those birds. If they did have high stress responses to this, they wouldn't have survived." Sapolsky suspects that all animals have innate mechanisms to cope with their expected environments, and that oxytocin may serve that purpose for mammal mothers, moderating expected intense levels of stress so that they don't interfere—or at least not too much—with mental function. "Somehow, mammals have worked this out, because cognition is a good thing to have when you have small dependents," he says.

There is still much to learn about oxytocin and the human brain, particularly because it's so hard to *study* oxytocin in the human brain. Although it's relatively easy to measure oxytocin in the blood, the hormone can't cross what's known as the *blood-brain barrier;* thus scientists still don't know whether blood and brain levels are comparable. A team of German scientists has recently been conducting experiments with nasal sprays that administer oxytocin to the brain. But in 2004, the spray was not available in the United States, meaning that until U.S. researchers can find a better way than spinal taps to measure brain chemicals it will be hard to recruit volunteers

Yet, the hardships involved in studying oxytocin may be just one reason the hormone's current celebrity has been so long in coming. Another may owe to the history of issues important to women's physiology, taking a back seat to those more important to men, who for decades have dictated the lines of research. One legacy of this bias may be enduring gaps in our understanding of the different ways men and women cope with stressful events.

The Costs of Stress, and Why Men Pay More

Ever since the early 1930s, the dominant model of stress response for humans and other animals has been the famous "fight-or-flight" syndrome first described by the Harvard physiologist Walter Cannon. Cannon showed how hormones help the body meet the kinds of emergencies faced by humans throughout most of our history, such as attacks by human or animal predators. What happens is that fear revs up the body and brain through the action of two separate systems. First, epinephrine (also known as adrenaline) and its cousin, norepinephrine, race through the bloodstream and out of nerve endings to make your heart beat faster and blood pressure rise. Circulation increases in your muscles, which presumably you're going to need. Your pupils expand. Your sympathetic nervous system has been activated, helping prepare you to hit the ground running by putting nonurgent functions, such as digestion, fat storage, and even your immune system, on hold. Next, your adrenal gland, just above the kidneys, releases glucocorticoids, which work over a longer period to back up the first-line emergency response.

Cannon coined the term *stress* to describe this process, and did so in approving terms, titling his 1932 book *The Wisdom of the Body.* Yet in the same decade, another scientist, Hans Selye, was beginning to conceptualize stress more in the manner that we view it today—as something that, in excess, can make you sick. Selye had discovered that his lab rats were developing peptic ulcers, enlarged adrenal glands, and shrunken immune tissues. He realized that he had been mistreating them unintentionally by handling them too much during experiments, and sometimes by accidentally dropping them; therefore, on a hunch, he began *intentionally* causing them discomfort by exposing them to frigid or hot temperatures, or forcing them to exercise. The rats became just as sick, leading Selye to conclude that physical stress was causing their illness.

Today, most scientists agree with the idea that too much worry—mental stress—and fear can also damage health. Yet only by the late 1990s did we begin to understand how differently men and women respond to these tolls. "Saying that what stress does to your body is preparing you to be a male on the savannah is ignoring half the data and is historically biased," says Stanford's Sapolsky, noting that dozens of scientific studies confirm there are "massive gender differences." One thing many behavioral studies show is that females, markedly more than males, respond to stress not by fighting or fleeing but by turning to others for mutual support.

This reaction calls more on the parasympathetic system, the "calm and connection" system in charge of those presumably nonurgent bodily functions such as growth and healing that are suppressed when you're gearing up for action. Prompted in part by oxytocin, it reduces the release of stress hormones, decreases blood pressure, and aids digestion. As Uvnas-Moberg writes, it inspires "friendliness instead of anger."

That females may be more inclined than males to handle stress in this way is an idea enthusiastically promulgated in recent years by Shelley E. Taylor, a psychology professor at the University of California in Los Angeles (UCLA). In a widely noted paper published in 2000, Taylor and five colleagues hypothesized that, compared to men, most women respond to stress less with the fight-or-flight model than with an alternate one she calls "tend-and-befriend." Taylor ties the abundant behavioral data to emerging research on oxytocin and endogenous opioids, and the apparently stronger role these mellow chemicals play in women's lives. At least one study has found that in rats oxytocin release in response to stress is greater in females than in males. Some experts believe this occurs because the female sex hormone estrogen enhances oxytocin's effects. "If you give oxytocin to female animals for five days in a row, they will have lowered blood pressure that lasts for three weeks," says Uvnas-Moberg. "In males, the effect lasts half that long."

In *The Tending Instinct,* Taylor's arguments about gender-related differences in reacting to stress refer to females in general. Yet it's clear that they're particularly relevant for mothers, because the "tending" in her formula refers to tending children.

Calming and Caregiving

Children are the main reason that the female model of stress response has evolved, Taylor suggests. Throughout most of human history, women have had to watch out not just for themselves but the small children they most likely had in tow. Thus, unless there was no other option, it didn't make sense for them to try to face down a predator. A mother who went on the attack might risk her life, leaving her offspring unprotected. Nor was fleeing a good choice, as anyone who has ever tried to hurry a toddler through a shopping mall can testify. No, for premodern mothers, the preferred tactic was to calm down her brood and fade into the scenery; in other words, tending.

Given men and women's historically different needs, it does seem reasonable to suppose that each gender has evolved differently in the matter of physiological responses to stress, Taylor argues—the male's to rev himself up, the woman's to quiet herself down. "If, as a mother, you flee from a menacing predator but leave your bewildered toddler unprotected, that child's chances of survival are clearly very poor," she writes. "Consequently, responses to stress that favored both the mother's and the child's survival would most likely be passed on."

In contemporary times, when the fear of corporate downsizing has replaced that of saber-toothed tigers, women's "tending" response to stress is still apparent, according to research by Rena Repetti, a clinical psychologist at UCLA who, as a full-time faculty member and mother of two daughters, has professionally pursued her understandable interest in the kinds of stress affecting working parents. In a unique study in the late 1990s, Repetti asked parents to report on how they treated their preadolescent kids after a stressful day at work—and then checked their responses by having one child in each family provide his or her version of events.

Repetti found that when fathers come home after a day during which there had been conflicts with coworkers and supervisors, both fathers and children reported that the dads were less responsive with their kids. Mostly, they withdrew. Mothers, by both mothers' and children's reports, became more responsive with their children after coming home from a stressful work day. They seemed more interested, affectionate, playful, and involved. They *tended.* "Having children or occupying the role of 'parent' may buffer at least some women from certain job stressors, and this result may suggest *how* this happens," Repetti concludes. "Increased positive involvement with children may reflect a coping process, one that can buffer parents from some of the ill-effects of job stressors."

When babies are involved, breastfeeding, as noted above, may be a particularly efficient way for mothers to quiet down. Before hearing about Repetti's work, Kathy Mayer, the pumping-and-driving internist mom mentioned in Chapter 5, was consciously using breastfeeding as a strategy to combat stress.

"If there's a voice-mail I've listened to as I'm driving home, and it makes me tense, I tell myself that I'm going to go home and breastfeed my kids and hang out with them, and answer that voice-mail later," she says. "The experience just totally calms me down and clears my mind if I'm strung out, and sometimes keeps me from saying something I'll regret."

Although many mothers become frustrated with nursing infants—the well-known Italian interviewer Oriana Fallaci called it "torture"—others report that once they master it, the experience becomes as enriching for the mind as meditation, slowing them down and giving them practice in being present in the here and now. Sue McDonald, the San Francisco mid-wife who skillfully helped steer me through my first pregnancy, recalls a childhood friend of hers who grew up to be a dedicated yet temperamental pianist. "She told me that her baby was such a gift," McDonald said. "She was nursing her one day, and whispering, 'Hurry up! I have to practice!' and then it struck her. Wait. Look at where I am. Look at what I'm doing. It gave her a huge new sense of perspective."

Social Chemistry

In addition to such caregiving closeness, another conspicuous feature of what Daniel Stern, the Geneva psychology professor, calls the "motherhood mindset" is the craving for a network with other experienced moms, which can supply both practical and psychological support. Overnight, your social world may be transformed. Your own mother and mother-in-law suddenly seem much more interesting, not only for everything they can teach you (Where's his soft spot, again? Does he sleep on his stomach or back? Shouldn't he be talking by now?) but also for how deeply they can understand what you're experiencing. You share a sudden strong bond with other mothers, too; but men, including your husband, may seem less enthralling. "We get put out to pasture," is the way one of my husband's old traveling buddies once put it.

In a study Stern conducted in Boston, he asked new mothers how much and what kinds of contact they had with others right after their babies were born, and was surprised by the "huge" amount of daily contact they had with other, more experienced, mothers. "In the average day, each new mother had more than ten different contacts, in the form of either visits or telephone calls, almost one for every waking hour," he writes.

Eleanor Bigelow, a forty-something mother of two children younger than three years, discovered her own circle of other mothers when she took off from work after several years at a high-pressure job as a senior vice president in a leading San Francisco brokerage firm. Previously, she had hung out mostly with men during the day, enjoying the macho camaraderie. Now her main social contacts are other mothers, whom she meets with their infants, at her home or theirs, to chat and compare notes. "I feel less crazy and less alone when I vent to another mother who appreciates my situation," she says. "Our discussions include family issues like how to raise kids, religion, schools, etc., the kinds of things single people don't dwell on. Because of this, I feel my friendships now are nurturing and more family oriented. I feel like we are all rowing in a boat and trying to help each other out. The mother network is strong and supportive."

Why quite so strong? Perhaps at no other time in a woman's life, other than being a child herself, or elderly and ill, does she depend so much on others; and, as the author Fay Weldon once noted, "weakness admitted is the very stuff of good friendships."

Weakness admitted is taboo in most workplace environments, yet a poignant refrain among overtaxed mothers. A shared awareness—of how great the stakes are, how much we're trying to do, and how often we're failing—binds us together uniquely. I simply can't imagine any men or single women carrying on a friendship quite like the one I have with Elizabeth Share, a fellow working mother of two young boys who lives about four blocks away, and whom I rarely ever see. Instead, we write e-mails and conduct bizarre attempts to talk on the phone. Here's a typical scene: I'm fixing dinner, my husband still at work; one kid is in the bathtub, the other at his computer. My computer will be on as well, in the laughable hope that in the few minutes when the kids might be occupied on their own, I can put in a bit of extra research. The boys are yelling, roughly every two minutes: "Mom! Get me the towel!" "Mom! I want juice!" "Mom! I spilled!" "Mom! I'm done with the juice!" "Mom!" "Mom!" "Mom!" "Mom!" I call Elizabeth and hear *her* two boys yelling in the background.

"I can't talk right now!" I shout.

"I can't either!" she shouts back. Then she calls ten minutes later and we go through the same routine. It's as if we're drowning, each coming up for air only to see the other's hands flailing in the water. Yet, somehow, it's richly comforting, and we go back to our respective challenges (she has her computer on, too, in the same laughable hope), just a bit fortified. For the rest of the evening, no mom is an island.

Strong female networks clearly help women become smarter and more efficient in their jobs as mothers. Once your kids start school, useful gossip about their teachers or playground bullies or who's signing up for which sports team circulates like strep in winter. When the networks are functioning at their best, you're not only trading gossip but favors and problem-solving advice, and even help in emergencies. Among our fellow primates, the difference this commingling can make is stark: In a sixteen-year study of wild baboons, published in 2003, the degree of sociability of adult females was shown to have made a positive difference in their infants' survival rates. That is, the more social the female baboon—the more time she spends close to other females, in mutually enjoyable pursuits such as picking twigs out of each other's hair—the more likely her baby is to survive its difficult first year of life. "Social animals are social for a really good reason," notes Susan Alberts, a biologist at Duke University who participated in the research.

The scientists theorized that the baboon mother's network created a positive environment for the babies, and shielded them from potential predators. For human mothers, this deep-in-our bones understanding of the difference that belonging to a community makes may explain why so many studies demonstrate that women's tendency to seek social support under stress ranks, as Taylor writes, along with giving birth as "One of the most reliable sex differences there are."

And, as with tending, the difference makes sense in the context of the division of labor that has lasted through most of human history. While the guys were out stalking mastodons, women had to cooperate more intimately, looking after each other's children, even as we do today, to give other moms a chance to watch for danger, look for berries, or just sit back and take a few deep restorative breaths.

As this behavior became a way of life, our brains may have evolved to reinforce it, by refining a delivery system of chemical treats that today keep women seeking intimate friendships with other women. We already know, from Jeffrey Lorberbaum's brain scans, that mothers, unlike fathers, respond to the sounds of their babies crying with activity in their "reward" centers—a reinforcement of caregiving activity. Something similar might be going on with female friendships, as suggested by some intriguing, although still unpublished, research presented at the annual meetings of the Society for Behavioral Medicine in 1999. In this study, Larry Jamner, a psychologist at the University of California at Irvine, and his colleagues gave a group of twenty-four women and twenty men tablets containing naltrexone, a chemical that blocks the level of pleasurable opioids circulating in the blood for at least twenty-four hours. The participants all kept diaries, which later revealed dramatic differences between the way the men and women reacted. The men's behavior remained essentially the same, but the women cut back on their social contacts, including phoning friends and spending time with them, reporting that they no longer felt the same pleasure in being close to others. The women, but not the men, noted that they also felt less alert.

Jamner's interest had grown out of evidence in animals that social behaviors result in the release of endogenous opiods, believed to reinforce sociability with a sensation of reward. His research raises the question of whether social ties may really be more innately rewarding for women than for men.

To date, we have only these sorts of slim pickings of evidence in the quest to understand the neurochemical underpinnings of women's bonds with each other. It simply hasn't been a major field of scientific study. But Barry Keverne, a professor of behavioral neuroscience at Cambridge University, in the United Kingdom, who has extensively researched the biochemistry of maternal behavior and social bonds in sheep and monkeys, writes that he is finding similarities between them: "Biology is incredibly conservative," he says. "If you have a mechanism that serves to help with mother-infant bonding, it's likely there on many occasions."

Trust Your Oxytocin

Is oxytocin involved in close human female friendships? At this writing, there's no direct evidence of this. Yet as researchers of many different disciplines have started to pay more attention to the hormone, they're producing some intriguing circumstantial findings and theories. Some scientists hypothesize that a deficiency of oxytocin in the brain might be a factor in autism, which is a genetic disorder that makes close relations with others impossible. And a remarkable experiment in 2003 demonstrated the close relationship, in healthy adults, between blood oxytocin levels, social bonds, and trust.

In that study, Paul Zak and colleagues recruited a group of volunteers, all unknown to each other, paid them each $10 for showing up, and then paired each one with one other person. The teams played a computer game in which one person in each pair had the chance to send none, some, or all of their money to his or her partner. They were told that whatever is sent will be tripled; that is, if the first player sends all of the $10, his partner will receive $30. The partner can then choose whether to send some money back, but is under no obligation to do so. Thus, the first person has the opportunity to send a signal of trust in this temporary relationship, and the second person may or may not be trustworthy.

After the players had made their decisions, Zak and his fellow researchers took blood samples to measure their oxytocin. They found that the players who received higher amounts of money, indicating trust, had higher levels of oxytocin in their blood, and were then more trustworthy themselves, as they shared more money with their partners. Zak said he was surprised by how strong the hormonal response had been, especially in the sterile lab environment, in which the game players were related only by computers. "The effect of oxytocin on face-to-face interactions must be quite strong," he noted.

Zak calls his avant-garde approach *neuroeconomics,* a discipline that focuses on the role of neural processes in financial decisionmaking. He believes that differences in trust, and oxytocin levels, across nations help explain their respective successes in living standards. And he argues that world leaders can and should improve those oxytocin levels by promoting breastfeeding, and in a sense, being more trustworthy themselves, by investing more in such feel-good endeavors as education and combating pollution. "Across the board, higher-trust countries have higher returns on their stock markets—and higher average incomes," Zak says.

In this grand scenario, trust, social bonds, and oxytocin seem like a smart combination. But I found an equally compelling example closer to home. In 1961, my parents and their four children moved to California from Minneapolis. It was my father's decision to uproot the family from the city where both my parents' parents and many childhood friends still lived—and it was my mother's job to form a new community. Her two sons were then aged seven and nine, and, as she told me years later, she determined they wouldn't be bar mitzvahed in an empty synagogue. And so she went to work, joining the temple's sisterhood, raising funds, driving carpools, and baking rich desserts to give as gifts. And soon she formed a wide circle of friends, most of whom were parents of her children's schoolmates. Not only were both bar mitzvahs well attended, but today, with her children in their forties and fifties, she still regularly sees and enjoys many of these same friends who responded to her need for support, even as we, her children, have all but forgotten the playmates of those days. Every time I see the statistic showing that women on average live more than seven years longer than men, I think of my mother—her all but entirely vicarious enjoyment of fattening foods, her religious exercising, but mostly her sociability.

Did her lifetime of close friendships help make my mother so calm and wise? Or is Uvnas-Moberg right in theorizing that the direct neurochemical experience of having four children changed her brain and somehow buffered it from stress? Or did both factors combine? We can only speculate about the connections. The evidence is already accumulating, however, that a tending-and-befriending style of stress reduction, and, possibly, the hormones that support it, is healthy for your brain and your body. Many studies have established that social ties help cut risks of stress-related disease by lowering blood pressure and heart rate. Indeed, the physical benefits of sociability have become so recognized in recent years that some doctors' offices now routinely ask new patients how often they see friends.

So sociability is smart, for mothers and others, when it comes to physical survival. Yet, even as we recognize mothers' extra incentive and drive to be social, cooperative, and trusting, it's important to remember that they can also be some of the most fiercely motivated and competitive creatures on earth.

CHAPTER

7

Motivation

The Mental Strength of Motherly Love

Because of deep love, one is courageous.

THE WAY OF LAO TZU,
SIXTH CENTURY B.C.

OLIVIA MORALES STANDS out among the mothers at her children's elementary school in a privileged suburb of San Francisco. Young, unmarried, and Mexican, she speaks only a few words of English and cleans hotel rooms all day long to pay her rent.

In 1999, while she was living with her parents in the Mexican border town of Mexicali, Morales gambled on changing her family's luck. Her children were then eight and six, and their father had abandoned the family several years earlier. Morales was earning $20 a week from her job on a computer monitor assembly line, and it wasn't enough to keep her children in school. And so she decided to borrow $1,500 to pay for falsified immigration documents, leave her children temporarily with her parents, and board a north-bound bus. "I had no luggage—just a little purse with $60 and some lipstick," she explains as she sits in the kitchen of the two-bedroom apartment she shares with another immigrant family. "And I was so scared of being robbed that I didn't sleep; I just held my purse like this—" she says, reenacting the scene with clutched hands and wide eyes.

All went well, however, and within three years, Morales had earned enough to pay for a *coyote* to smuggle her children across the border.

"They'll speak two languages here, and they'll learn to use computers," she says. "They'll have better futures and more possibilities than I did."

Morales is far from alone in her willingness to take such an exceptional risk for her children. During the past three decades, thousands of Mexican, Central American, and Caribbean women have boldly faced these same demons when crossing the U.S. border to improve their families' lives. Their journeys are just one illustration of how motherhood can be a major motivator.

Motivation is a key component of *emotional intelligence,* as defined by Peter Salovey, the Yale University psychologist who coined the now-famous term in 1990. Daniel Goleman, author of the best-selling book *Emotional Intelligence,* calls "positive motivation" the "master aptitude"—one he defines as an ability to marshal enthusiasm and confidence that's held in common by Olympic athletes, world-class musicians, and chess masters. In mammals, a mother's longing to be near and fend for her children may be the strongest motivator of all. In a 1930 experiment, scientists separated mother rats from their litters with an electrified grid. They found that the mothers were more willing to endure shocks in order to return to their babies than other rats, deprived of food, drink or sex, were willing to take to reach the object of their desires.

Deep in our genes, human mothers are subject to a similarly powerful drive to be with their babies, nurture them, and keep them safe. "The reason mothers do all these generous and brave things is not because it's necessarily good for the mothers, but because *they* had mothers who did the same, and *because* those mothers did those things, they had more offspring surviving, with the same system of genes responsible for that behavior," points out Randolph Nesse, an evolutionary theorist at the University of Michigan. We share this operating system with other mammals, and it appears to have the power to transform even our much more sophisticated lives. Long after the hormones of attachment have subsided, many human mothers find themselves to be more disciplined, fearless, and, in new ways, ambitious, all amounting to a very elemental kind of smarts.

"A baby is a constant mission," says John Ratey, the Harvard psychiatrist, who adds that this kind of sense of purpose can improve someone's focus and ability to screen out distractions. Ratey, an expert in attention deficit problems, said that one woman he treated realized she had an attention problem only after her baby had turned three and was in day care. "This was when she realized something was different," he said. "She'd gotten used to having a mission. Eventually, she decided to go to back to school and get her Ph.D."

"Task Ownership" and Responsibility

Tanja MacKenzie, a "troubled teen," was, as she says, "always getting into mischief and staying out late, not listening to [her] parents." She drank, smoked dope, and says that she had such low esteem that she felt "vulnerable to boys." Then, at the age of sixteen, MacKenzie became pregnant and turned her life around. Living at a home for teenaged mothers, she devoted herself to her baby for two years while taking mandatory parenting classes. Then she went back to school part-time, and eventually she started her own business: sewing clothes at home so that she would have more time mothering. Becoming pregnant as a teen was "the best thing that could have happened to me," says MacKenzie, who now lives in Ontario, Canada, and is married and the mother three other children. "It made me take responsibility for myself and be so much more conscious of my actions and how they affect others. . . . Having the responsibility for another person gave my life a sense of purpose."

MacKenzie's success, which had a lot to do with a remarkable degree of institutional aid, sadly isn't more the rule. Teen pregnancy remains a serious problem because a high number of young mothers, especially those in poverty and lacking social support, become so negligent that, as a group, their babies face lower odds of survival. Yet MacKenzie's determination is more common than many people assume. "We hear the phrase 'teen parents' and think the worst," says Sue Hagedorn, who, as head of the School of Nursing at the University of Colorado, has observed hundreds of young mothers in the past decade. "But I've seen most of them get more organized in the process and really do a good job. Motherhood seems to give them a sense of purpose they wouldn't otherwise have had."

Indeed, for many women, motherhood provides an introductory experience of having real power in the world—power enough to create and defend a new life. And with great power, as any Spiderman fan knows, comes great responsibility.

Major auto insurance firms recognize the tie between having to care for a baby or child and being more self-disciplined. In California, State Farm Insurance, among other leading companies, offers large premium reductions for drivers who are married and for drivers who, married or not, have custody of a minor child. The assumption is that the obligation to watch out for a child makes the policyholders more careful. And although it may be an old truism that babies encourage their parents to "settle down" and become more mature, only in recent years, with the mechanics of motherhood under more study, have academic researchers found evidence of it.

In the mid-1990s, while interviewing college-student mothers about the social and economic support they relied on, Pam Sandoval, an education professor at Indiana University Northwest spoke with twenty-eight women who had given birth while in high school or college, and were told by most of them that they saw motherhood as an organizing force in their lives. "Before I had my baby, I was not responsible at all," said one interviewee, an unmarried African American woman who became pregnant between high school and college. "All I thought about was music and myself. When I had my daughter, it was like I have another person that looks up to me. I have to be responsible—that is it—no more games. Having my baby made me go to school. . . . I want to be able to say to my baby, 'Hey look, I got my degree so I can take care of you.'"

Sharon Hays, a professor of sociology at the University of Virginia, has collected similar testimonies from low-income urban women. "For many of them, prostitution and selling drugs weren't a problem—until they became mothers," she says. But once children arrive, Hays says, moral issues loom larger—perhaps in part because the mothers have more self-respect. "They say it makes them better people, more grown-up," she says. "And the fact is, for low-income women, mothering is the central fulfilling thing in their lives, there's just no question. If you want stimulation, you want to feel better about yourself, you want to feel you're doing something for the world, your job at Dunkin' Donuts is not going to do it."

The enhanced self-esteem that caring for a baby can bring to a mother probably has a lot to do with practice in being the boss. Daniel Stern, the Geneva psychologist who has specialized in the study of mothers, refers to this practice as "task ownership."

"The old expression 'the buck stops here' takes on a new meaning," Stern writes in *The Birth of a Mother.* "You will have to make split-second decisions even when you don't really know what to do and have never been there before. It's akin to being a CEO, a policeman on duty, or a physician on call. All eyes turn to the person in authority and expect that person to know what to do. . . . You own the responsibility so that any successes and failures, even if they are brought about by others, revert to you."

If the successes outnumber the failures, a mother may find herself feeling more competent in general, and consequently willing to take on new challenges. That's how it was for Lori Willis, thirty-three, who says she discovered new reserves of strength after her husband divorced her, following the birth of their son. "It's amazing how your life takes a turn for the worse and you're able to be strong for your child," says Willis, who rose to the challenge by applying for a major new job as an administrative coordinator of a Harvard Medical School psychology training program, where she negotiated a work week of ten hours a day, four days a week, to give her an extra day with her toddler. "You become more mature when you have a child. You take life more seriously," she says.

Ravenna Helson, the psychologist at the University of California at Berkeley, says that in the Mills College graduates she has studied for the past four decades, the women's quality of experience as mothers was definitely linked to changes in their personalities over time. If they felt good about their parenting, they tended to become more flexible and resourceful, less fearful, and more "dominant"—meaning focused and confident—in other realms of their lives. But if their mothering experience was negative, they eventually declined in these emotional strengths.

No Fear

When Susan Galleymore's twenty-six-year-old son, Nick, a U.S. Army Ranger, was transferred to Iraq in 2004, she initially stayed up at night, haunted by visions of him being maimed and killed. But she soon tired of feeling so helpless and resolved instead to travel thousands of miles on her own, and ultimately navigate a war zone, to check up on him. She arrived at his base with a box filled with See's Candy and PowerBars. "Hey, Nick, your *mom's* here," one of her son's fellow soldiers called out. Galleymore, who has since decided to channel her nervous energy into writing a book about wartime parents, says of her trip: "My son is there, and it's up to me to know what he's going to be facing and what's going to happen in his life."

Galleymore's bravery was as stereotypical as it was remarkable. Like the story of Olivia Morales and the mother lab rat crossing the electric grid, it involved not just a journey and a reunion but a mastery over fear. In the earliest stages of the mother-child relationship, this capacity, much like others involved in a mother's changed brain, appears to involve hormones as well as experience. And, as usual, it is most starkly evident in rats.

Timid by nature, rats generally prefer to hug walls and loiter in dark corners than to venture into open areas, where they might be caught by a fox or a hawk. But mother rats, breadwinners all, are obliged to think outside the box to provide food for themselves and their litters, a duty that can drive them to explore new areas and travel farther from home. With the onset of motherhood, rats demonstrate what seems to be a marked reduction of fear, a change which in their circumstances can translate to smarts, because by taking more risks, within reason, they can help more of their pups survive.

To test this hypothesis, Craig Kinsley and Kelly Lambert, the Virginia researchers, placed three groups of female rats—virgins, pregnant rats, and new mothers nursing litters—in the middle of a circular enclosure to measure how long they would stay in such an uncomfortable position. The virgins stayed in the open for only about five seconds before heading for the comfort of the wall, the pregnant ones lingered somewhat longer, but the nursing rats stayed longest of all—about a hundred seconds on average. The pregnant and nursing females also tended to freeze in fear much less often, and explored the area with what seemed like more confidence. They stood on their hind legs more often, so that they could survey their surroundings, and more frequently walked over little blocks placed in their way. "I've experienced this sort of thing in my career," says Lambert, with a laugh. "Like, who the heck cares? I'm going to do this thing."

Seeking clues about what might be happening in the rats' brains to make them so bold, Kinsley and Lambert performed an additional experiment. They subjected a group of rats to intense stress by restraining them in a clear, Plexiglas tube in a harshly lit room, after which the researchers "sacrificed" the animals and dissected their brains. Checking the rats' "fear centers," including the amygdala and parts of the hippocampus, Kinsley and Lambert looked for a protein, known as c-fos, that is produced when brain cells are active. There they found a clear difference between mothers and nonmothers. The mother rats' fear centers had been much less engaged.

Though much remains to be discovered about this transition from timid to bold, two mind-altering hormones of motherhood appear to be the main instruments of the transition. Oxytocin, as described in Chapter 6, may keep stress at bay while at the same time another hormone, prolactin, dampens anxiety and fear.

Dubbed "the parenting hormone," prolactin is named for its role in lactation. It rises in pregnancy and is further elevated each time a baby suckles, stimulation that sends "you-go-girl" signals to the mother's hypothalamus and pituitary gland. In lactating women, prolactin secreted in the blood rises to as much as eight times normal levels. Meanwhile, prolactin also is active in the brain, because, like oxytocin, it is also a neurotransmitter. "This hormone is fascinating," says Inga Neumann, a neurobiologist at the University of Regensburg in Germany, who has participated in some of the only prolactin research involving humans. "In the blood, it governs milk ejection, but in the brain it affects behavior, making animals braver, even daring to risk their lives."

Prolactin was first identified in the early 1930s by a scientist named Oscar Riddle, who found that injecting it into pigeons would make them "broody," that is, inclined to hover over their eggs. Since then, other researchers have found that prolactin also appears to make birds daring. High levels of the hormone have been found in birds engaging in a particularly bold parenting behavior: When a predator approaches the nest, either the mother or the father, or both, will turn themselves into a decoy, floundering on the ground, feigning an injury such as a broken wing, to lure the intruder toward them and away from the eggs.

Prolactin works wonders with rats as well. When researchers injected it into the brains of virgin females, the rats, like the fearless mothers in Kinsley and Lambert's experiment, became more willing to explore brightly lit parts of a maze. During pregnancy, the hormone's impact on a virgin female's normal timidity plays an especially vital role, because rats' innate fear of the unfamiliar usually includes little pups. As previously described, a female rat that has not been pregnant or, alternatively, given time to adapt to wee ones, will just as soon try to eat them or bury them as protect them. But once a rat has been pregnant, this fear disappears, allowing the female to morph from a potential pup-eater into a warm-hearted matron. Moreover, once a rat has undergone this hormonal change and actively mothered a litter, she will, for a long time to come, be much quicker to respond in a motherly way to other mothers' pups. This phenomenon, called "maternal memory," appears to be present in primates as well and provides strong evidence that a long-lasting change in the brain has occurred.

At this point, it's worth stating the obvious: Fearlessness isn't always smart. Lactating wild rats are over-represented in the traps that university researchers set in urban alleys, precisely because, like *Star Trek*'s crew, they tend to boldly go where no rats have dared to go before. Galleymore's journey to Iraq might also have ended badly, and the same goes for Olivia Morales. Kathryn Rodriguez, a human rights worker keeping watch on the U.S.-Mexican border in Tucson says that many would-be immigrant mothers fail tragically: "We're seeing more and more mothers crossing," she reports, "and more and more, they're dying in the desert."

All the same, for every rat mom stuck in a trap, others have thrived and multiplied by finding new sources of food. At last report, Galleymore was safely back at home, working on her book, in Alameda, California, and Morales's kids were learning English and doing their homework on computers.

Another Mother for War

Beyond plain fearlessness, mammal mothers are actually more famous for a different kind of courage—aggression against intruders. Scientific literature abounds with stories of rodents and primates unleashing their Inner Hulk in the face of threats to their offspring. Children are warned never to come between a mother cat or dog—even a formerly tame pet—and her babies. And this is only reasonable, focused as the new mothers are on bringing those offspring to maturity.

Most animals need to be vigilant about strangers. Male rodents and primates have a nasty habit of killing babies sired by other males while those babies are still nursing. With no further suckling to stimulate milk production, the mother begins to ovulate again, giving the killer a chance to sire his own offspring. Rodent moms also face threats from other females, who might kill pups in the process of competing for safe nesting sites, a scenario vaguely reminiscent of contemporary bidding wars for housing near good school systems in America's suburbs.

The anthropologist Sarah Hrdy was able to observe maternal aggression close up one afternoon after her son called her at home to ask that she feed his eighteen-inch-long pet ball python. Hrdy fetched a live, wild mouse from a trap under her kitchen stove and put it in the python's cage. An hour later, she returned to find the mouse, transformed into a furry

Ninja warrior, facing down an apparently terrified snake, which was huddling in a corner. Hrdy didn't understand just how the mouse had done that, but decided it had earned its freedom, and took it outside to release it. Only after her son returned and cleaned the snake's cage did he discover the reason for the mouse's heroism: Once saved from the trap, she'd given birth to a litter of pups, hidden away in the snake's lair.

Scientists and untrained observers alike seem endlessly fascinated by these stories, perhaps because they contradict the cultural stereotype of a peaceful Madonna. Many of us cling to this cliché despite overwhelming evidence to the contrary. Barbara Bush, Hillary Clinton, and Margaret Thatcher may seem to be exaggerations, but they are not necessarily exceptions when it comes to mothers prepared to defend their turf. The reality, maintains Sara Ruddick, who coined the term *maternal thinking*, is that a mother is constantly immersed in conflict, with her children as well as with an "outside" world at odds with her or their interests.

In humans, this more aggressive behavior can be most noticeable in the first few days postpartum, when husbands are most frequently the special targets of a new mother's wrath. In Toronto, Alison Fleming has documented evidence that women commonly undergo a postpartum decline in "positive feelings" toward their spouses in the first several weeks after birth. For many mothers, this is a huge understatement. It may seem that their marriages have overnight become zero-sum games. His nap is now your wakefulness. His career achievements are supported by your sacrifice. And so on. Yet as many a suffering spouse has protested, hormones may be boosting a mother's wrath. One small but provocative study by a group of researchers at Italy's University of Padova directly implicates high levels of prolactin in postpartum feelings of hostility.

As women discover just how angry they can be in those first hormone-powered days of motherhood, and revisit that state in days and years to come, they might find to their surprise that anger can sometimes be quite effective—even smart. Motherhood, in short, is powerful assertiveness training. Even the most timid woman may soon realize that her life will quickly go downhill if she doesn't establish some authority over a recalcitrant toddler. And even normally passive women, for the sake of their children, may force themselves to crack down and oblige their children to have measles shots and arrive at school on time. Once you start out on this road and become practiced at speaking your mind and getting your way, it's often hard not to keep on going.

In animals, maternal aggression rarely lasts very long. Most intense during nursing, it's rarely observed in a mother when her babies aren't nearby. Yet in humans, who keep their children close for so much longer, some vestige of new ferocity may linger long after the children are weaned. Julie Suhr, the Ohio State psychologist, recalls an afternoon when she and a few other parents were talking in one of their front yards while their children were playing some distance away. A stranger approached the children to tell them he'd lost his dog, but hardly got the words out before, as Suhr recounts: "Wham! In a split second, we were all out of our conversations and right over there. . . . This guy couldn't understand why all these parents were marching on him until one of the mothers said, 'Why the heck would you go over to the kids when the parents are right here?'"

A mother's feistiness can take many forms, including willingness to stand up against society and authority. Try a Google search with the phrase "mothers against," and you'll come up with scores of listings. There are mothers organized against violence, drunk drivers, and video-game addiction. There are other mothers against gang wars, guns, and fathers in arrears. There are Mothers Against the Death Penalty in Uzbekistan and Mothers against GMOs in New Zealand. Still other mothers are taking on poverty, war, dioxin, circumcision, methamphetamine, and allegations of Munchausen Syndrome by Proxy. I also found a listing for what appears to be an apocryphal group called Mothers Against Peeing Standing Up.

In my travels through Latin America in the early 1990s, I witnessed one of the boldest mothers' groups of all in the Mothers of the Plaza de Mayo, in Argentina. The group was formed by scores of women, from wealthy and low-income families alike, whose sons and daughters had been "disappeared" by soldiers and right-wing death squads during the military dictatorship in that South American country from 1976 through 1983. (Ultimately, as many as 30,000 Argentines were killed on the spot or detained in the so-called Dirty War, never to be seen again.) For years, these mothers suffered the agony of not knowing whether their children were dead or alive. At first, they sought answers in traditional ways, such as visiting police stations or hiring lawyers. Then they took the daring step, for those repressive times, of organizing to bring international pressure against the evasive military leaders. Every Thursday afternoon, dressed in black with white head-scarves, the Mothers began marching around the plaza in view of the presidential office. They soon became the most visible and powerful challengers to the dictatorship. "We'd grown tired of knocking on doors and going to every possible place . . . tired of feeling defrauded, deserted and marginalized," the group's president, Hebe de Bonafini, has explained.

In time, the Mothers surpassed their original aim of discovering their own children's fates, their mission broadening to defend all of the country's human rights victims. They braved death threats and eventually also the "disappearances" of some of their own members. And once a form of democracy returned, with a new government offering a controversial pardon for the military officers, the Mothers kept marching in protest. By the time I first met their members, their numbers had declined, yet dozens were still keeping up their weekly vigils, railing against everything from government corruption to police abuse, their determination seemingly undimmed.

Evolving Ambition

Like few other experiences, motherhood can make a woman re-examine her priorities and re-direct her energies. Some decide to step back from a demanding professional career. Some discover new passion for their work. And some seek new ways of making their professional work more congruent with their lives as mothers. Yet although the Mommy Brain cliché might lead us to believe that women generally become retiring, more "nurturing," and less ambitious as mothers, there is plenty of evidence to show that we become even more competitive.

"From the moment Chelsea was born, my ambition focused on me and my mortality," says Rayona Sharpnack. "I was already thinking out at least two generations, to her and her children, which raised the stakes as far as what I needed to get done."

Sharpnack, previously a professional athlete, a junior high school teacher, and business consultant, founded her own business, the Institute for Women's Leadership in Redwood City, when her daughter was four. She has since coached hundreds of professional women. And her gung-ho attitude about her work—while she's just as gung-ho about mothering—is completely unsurprising when viewed in our evolutionary context, according to Sarah Hrdy.

For most of human history, a mother's ambition was "an integral part of producing offspring who survived and prospered," Hrdy notes. "Striving for local clout was genetically programmed into the psyches of female primates during a distant past when status and motherhood were totally convergent." Thus, mothers are just as likely as other women or males to seek status in the fields that matter to them. Among primates, including marmosets, tamarins, and some baboons, this tendency shows itself when dominant females harass subordinates, sometimes leading the poor lesser apes to delay ovulation or spontaneously abort. And in South American tribal societies, mothers aggressively seek lovers who become "secondary fathers" to their children, providing extra food and other comforts that help them survive.

An inescapable reality today is that many millions of mothers in modern Western cultures, no matter how much they love their children and want to be with them, have little choice but to strive for clout in a competitive job market in which their paychecks stand between their children and poverty. The 2000 U.S. Census found a record 10 million single mothers living with children younger than eighteen, up from 3 million in 1970—implying a greater financial responsibility for mothers than ever before. Even so, every few years since women entered the marketplace in force, there have been prominent media "trend" stories about mothers abandoning ambition and heading back home.

Most, if not all, the women featured in these articles belong to an elite minority relying either on their families or their husbands for financial support. When the *New York Times* published a front page story in 1980, headlined "Many Young Women Now Say They'd Pick Family Over Career," the interviewees were Ivy League undergraduates who, despite their complaints, were going off to medical school and other graduate studies. Then, in 1986, a cover story in *Fortune* magazine (followed by copycat stories in *Forbes, USA Today,* and other publications) promised to explain "Why Women Are Bailing Out." The story claimed that "after ten years, significantly more women than men dropped off the management track." Yet, as the writer Susan Faludi recounts in her book *Backlash: The Undeclared War Against American Women,* when she confronted the article's author with evidence that the trend he'd reported didn't exist, he conceded that the drop-out rates for men and women were roughly the same.

The periodic reports of the death of women's workplace ambitions nonetheless continued, each time evoking much attention and debate. A *New York Times Magazine* cover story in 2003 about the "Opt-out Revolution" is a case in point. Faithful to tradition, it zeroed in on a small group of successful, well-educated women who had chosen what one of them bluntly called the "escape hatch" of maternity. And this story, too, inspired others, including a *Time* magazine cover titled "The Case for Staying Home: Why More Young Moms Are Opting Out of the Rat Race." Yet the main statistical hook for this new round of stories was the fact that work force participation of married women with infants had fallen from 59 percent in 1998 to 55 percent in 2000.

Considering that in those same years a recession had made it harder for mothers, and others, to hold onto their jobs, it may have been a stretch to declare that a revolution. And again, the decrease was concentrated among women who could most likely afford to take off, including older and more educated mothers. Among African American mothers, and those who had not graduated high school, the percentage of working mothers with infants slightly increased during that time.

Moreover, the "retiring to the home" of energetic and resourceful females who can afford to do so is often a matter or redirected rather than reduced ambition, as the *New Yorker* so well described in an article titled "Mom Overboard! What Do Power Women Who Decide to Quit the Fast

Lane Do with Themselves All Day?" It focused on women such as Sera, a "conspicuous achiever" who had given birth to four children spread over six years and enrolled them in classes in dancing for toddlers, specialized jazz, eurhythmics, gymnastics, drawing, arts, chess, sailing, and Spanish.

"Opting out" of the formal workplace can clearly thus lead to just as competitive and ambitious a life, if not more so, than before. Still, for every privileged mother who takes that route, many others, like Sharpnack, manage to bring new resolve and energy to their work lives.

One of these now calls herself Vaschelle, no last name. She walked out on the physically abusive husband she had married at nineteen and took her three-year-old daughter to a small island in Hawaii, where Vaschelle became a counselor for victims of domestic violence, started a thriving massage business, and eventually trained herself as a freelance photographer. "I had to make something of myself for my daughter's sake," she says. "We couldn't count on anyone else."

For other women, already dedicated to their careers by the time they become mothers, the new commitment at home can give more meaning to their work. "I want my kids to be proud of me, so we talk all the time about what I contribute on the job," says Rhonda Staudt, an engineer and mother of three in upstate New York. One of Staudt's greatest contributions came in 1999, when she led an innovative team at Xerox that managed to create a waste-free manufacturing facility. That year, as she recalls, a team of executives from other corporations toured her plant. Their guide began explaining that the accomplishment was a result of how much engineers love to rise to difficult challenges.

"I was rolling my eyes, thinking: Okay, that's why *you* guys did it," Staudt says. Yet when the guide turned to her and asked why she had pushed so hard to meet the environmentally benign goal, her tough-guy composure dissolved and she burst into tears. "I am a mom," she said, simply.

"It was like they were asking, what is the *essence* of you?" she elaborated, in recounting the scene. "I'm a mom. Long before I was a mom, I knew I'd be a mom. And I want my kids to grow up in a good place, and to be responsible—not just responsible to family and religion but in the way they live, and the impact they have on their world."

Indeed, if you look at the membership of the fast-proliferating environmental movement today, it's crowded with working mothers. One obvious reason is that nonprofits are usually friendlier and more flexible than the rest of the working world. But another appears to be the pull of idealism. "Women and moms are taking over this movement," says George Basile, a senior scientist at The Natural Step, a San Francisco group that advises corporations on how to tread more lightly on the Earth. "They just have a different sensibility, and maybe that's what it's going to take to finally make some real change."

For many women, the desire to spend time with their own children—often combined by the more future-oriented attitude about work that motherhood can bring—inspires creativity in their careers. Confronting the change in their own lives, they boldly insist on corresponding changes in their work.

Such was the case with Hrdy, who as the mother of three children switched from doing fieldwork on primates, which required constant traveling, to teaching and pursuing U.S.-based research on humans, which allowed her to be home in time for dinner. In the process, she wrote two highly influential books that have solidified her reputation. Yet some of her colleagues, as she notes, still speak in terms of her having "dropped out" of academia.

Anne Gelbspan went through a longer and more difficult transition after having her children. Born in Sweden, she'd been used to a life of travel and adventure. She ran a business marketing Middle Eastern fabrics, hitchhiking in Tehran, and living for a time in a North African village. Then she met her future husband, the investigative journalist Ross Gelbspan, and ended up buying a one-way ticket to New York City. She gave birth to two daughters within two and a half years, after which she found herself feeling trapped in the family's small urban apartment while her spouse worked long hours away from home. To top it off, both her babies had major sleep problems.

"I became clinically depressed," Gelbspan recounts. "I hadn't really had a clue about what I was getting into." She felt her only refuge was to leave the apartment with her babies and head to what she came to think of as the "watering hole," the playground in Central Park. There, she could share childcare and socialize with other mothers, exchanging information on sleep remedies, diets, and sitters. The sense of how vital it was to have such a meeting place stuck with her as she gradually became "unclogged," as she describes it, and eventually she used that idea to transform her life. She enrolled in classes in environmental psychology, and then, after she and family resettled in Boston, she looked at her new neighborhood with new eyes. "We had a great setup, with a playground right near our apartment, but we were five minutes from Roxbury, a lower-income part of town where there wasn't anything like that," she said. "I know that before I became a mother, I wouldn't have seen the world in that way. I didn't have such a need for community; I'd always felt very independent."

Gelbspan's inspiration eventually led to a career as a nonprofit developer of low-income housing with Boston's Women's Institute for Housing and Economic Development. There, she combines the architectural sensibilities from her youth in Sweden, where, as she says, "the common good is an accepted thing," with her entrepreneurial drive and her insights as a mother. Her daughters have grown up to share her social conscience, one working for the charitable group Oxfam and the other studying finance and corporate responsibility.

Napoleon Bonaparte once wrote of "two o'clock in the morning courage," which he defined as "unprepared courage," the courage of someone who is improvising. He was thinking of warriors, not mothers, yet the image suits both. Filled with the great love that inspires courage, many mothers learn to adapt quickly to changing circumstances, taking previously unimaginable risks, and in the practice, becoming ever bolder. Yet even as experience teaches mothers the value of motivation and assertiveness, we'll most likely find ourselves more routinely engaged in other forms of social intelligence, including empathy and optimistic spin control.

CHAPTER

8

Emotional Intelligence

How Motherhood Teaches Social Smarts

Imagining motherhood opens the door to imagining every power relationship, every profound connection. . . . Far from depriving me of thought, motherhood gave me new and startling things to think about and the motivation to do the hard work of thinking.

JANE SMILEY,
CAN MOTHERS THINK?

IN THE MID-1990S, at a neuroscience lab in Madison, Wisconsin, South Asian Buddhist monks underwent a series of brain scans to give U.S. researchers insights into their considerable ability to cultivate kindness. These "Olympic athletes . . . of meditation," as Richard Davidson, a psychologist at the University of Wisconsin, calls them, spend hours each day in contemplative exercises, such as imagining events that make them angry and then transforming their experiences into compassion. Their dedicated practice makes them masters of calm. And they come to mind as I listen to the screams of my son Joey, whom I've just denied a second dessert. He is using a phrase *I* sure didn't know by third grade, and furthermore wouldn't ever have dreamed of applying to my mother. He doesn't seem at all cute just now, and I'm aware that some part of my brain is preparing a counterattack, a shouted rendition of every sacrifice I've ever made for him, from cutting down on coffee while pregnant to rising before 7:00 A.M. last Sunday.

We lock eyes. I take a breath. And then I tell Joey that I love him; I see he's upset, but his language isn't acceptable, and he needs to think before he acts. I mentally picture a convention of parental advice gurus giving me a standing ovation as I hope my controlled, mature example of not screaming my own favorite obscenity will serve Joey somehow, someday. I certainly don't always succeed at this trick. But today, I sense I've made progress in my efforts to improve my "emotional intelligence," a capacity that has been studied and prized more in the past two decades than during any previous time.

The idea that we have "multiple intelligences" going beyond IQ was introduced by Howard Gardner, a psychologist at Harvard, in 1983. His list of seven types of intelligence ascends from the capacity to understand words and numbers—linguistic and logical intelligences—to the capacity to understand yourself and others: *intra*personal and *inter*personal intelligences. Seven years later, Peter Salovey, a psychologist at Yale, and his colleague, John Mayer, coined the term *emotional intelligence,* defining it as encompassing "the ability to monitor one's own and others' feelings and emotions, to discriminate among them, and to use this information to guide one's thinking and action." These sorts of capacities, which used to be referred to as "good character," can contribute to stronger friendships and marriages and, apparently, even better physical health, all making for a happier life. There is also evidence that they may provide a competitive edge in the workplace, especially in jobs involving a lot of personal contact, such as teaching, medicine, and sales. They all seem to improve with practice, and there are few more intense and repetitive opportunities for practice than everyday parenting.

As it turns out, Davidson, a colleague of the brain-scanner Jack Nitschke, who was cited in Chapter 3, is working with mothers as well as with monks. Both groups—monks consciously, mothers often *un*consciously—routinely exercise positive emotions, such as love and compassion, which underlie emotional intelligence, and, according to some experts, can be summoned more easily with experience. "Love as well as other positive emotions are not static but can be learned as skills," contends Davidson. "They're skills that are not dissimilar from the skills that you might learn riding a bicycle or learning some other kind of complex motor behavior. If you train them, they will increase in strength and frequency."

For just this reason, emotional intelligence, or "EQ," is probably the main arena in which motherhood can make you smart. Your love for your child can uniquely commit you to your practice. Moreover, in rearing children, you have constant opportunities to summon empathy, a basic ingredient of EQ, while cultivating emotionally smart techniques that include self-restraint, conflict resolution, and "reappraisal," which is neuroscience parlance for the spin-control we practice when we reconsider negative impressions in a more positive light.

Chains of Love

A mother's love is often described as one of the purest of emotions. Yet a more analytical perspective suggests that a mix of selfishness with selflessness, and hormones with experience, is what ties us to our children so fiercely: This potent and complex force is what makes us work so hard, and so inspires us, to learn how to manipulate them for their own good. At the heart of mother love is a desire for the closest thing to immortality that a human being can achieve. An ancient and mostly blind part of me seeks to have my genes replicate themselves, Joey being their vehicle. And this replication will be less likely if Joey tries out the phrase he just screamed at me on a potential mate or on a hot-headed highway patrolman.

This evolutionary pull, assisted by a scheme of hormones of attachment and neurotransmitters of pleasurable reward, works wonders in binding mothers to their babies from the first besotted days when you both still feel like one person. In one group of indigenous Bolivians, both parents change their name with the birth of each infant, adopting a version of the new baby's name. Among U.S. parents, a survey of couples with babies aged three months showed that 73 percent of mothers and 66 percent of fathers thought their infants to be "perfect." I look at my own postpartum journals about Joey and find embarrassingly rapturous observations, such as: "His long fingers are a sure sign of genius!"

James Leckman, the Yale researcher whose work was discussed in Chapter 3, is fascinated by this phenomenon, and in his brain scans of mothers he has been seeking confirmation of the nexus that he is certain exists between parental love and mystical experience. (Previous *f*MRIs suggest that spiritual experiences involve a network including the amygdala, the parietal lobes, and the right frontal cortex, which has been called a "God module" in the brain.) As Leckman points out, religion and parenting both involve idealization, a transformation of what is thought to be most important in life, an exclusivity of focus, a longing for reciprocity, and the feeling of experiencing an event impossible to describe.

"Look at it this way," Leckman says. "We've all had this experience of being helpless and hungry and having someone much bigger than ourselves appear and take care of us. Otherwise we wouldn't be here today. And all of this happened before we had words to express it." These powerful feelings help explain the remarkable sense of fulfillment that keeps most parents committed to their children through the sleepless nights of infancy, the tiring trials of youth, and the exhausting crises of adolescence. Most likely, at least at first, they involve the brain's "reward" circuit, which, as described in Chapter 3, activates in apparent pleasurable expectation when mothers look at their children or even listen to their babies cry. In a remarkable experiment involving a miniature *f*MRI device, Craig Ferris, a psychiatrist at University of Massachusetts Medical Center, has found that the reward circuit is also activated when mother rats nurse their pups. "Behavior that activates the reward system is likely to be repeated, since reward is a positive reinforcer," Ferris says. "Activation of a pleasure sensation in the mother may be part of the reason that rat mothers bond with their pups, and this may also occur in other species."

Michael Merzenich, the brain plasticity expert in San Fransisco, believes that in humans, too, breastfeeding—or even what he politely calls "ventral-to-ventral contact," the prolonged, belly-to-belliness between mothers and infants—can provide much of the glue in a potentially lifelong bond. He describes breastfeeding as a kind of temporary breakdown of identity for both mother and child that powerfully affects both their brains. "At that point, the baby and mother are unified, something that simply has to contribute to a mother's idea of herself," he says. "If you want to think about where empathy comes from, it comes from that interaction. And it absolutely can extend to other people. . . . It's a heavy dose that unfortunately a male could never experience in the same way, and maybe a source of how women think about people in the world differently than men."

Emotional Intelligence 101

The repetitive, intimate experiences, such as cuddling, feeding, and changing diapers, that most mothers have with their babies can amount to a crash course in how someone very unlike you reacts from moment to moment. To mother an infant is also to understand how profoundly your own pleasure, or mere peace, can depend upon the welfare of another human being, and consequently, to learn with great urgency how to understand and cope with someone else's feelings.

This new appreciation dawns on you with special power, coming as it does at a time when you're probably more physically and emotionally vulnerable than you've ever been. The act of giving birth is many a woman's first real experience with utter loss of control. Bowled over by morning sickness, you can begin to imagine what it's like to be on chemotherapy. Pushed around in wheelchairs, you get a preview of what it's like to be old. Like a cult convert, ripped from the familiar trappings of your previous life, you drop your defenses and find yourself more open to the influence of others.

In the process, you may find that you're "resetting your hedonic homeostasis," as Leckman describes it: Most of what you do, or fail to do, will influence your new baby, who will quite likely in time influence you in return. "All of a sudden, you have to be tuned into the feelings of someone else in an incredibly nuanced way," says Salovey. "It's intense training with strong reinforcements."

What constitutes these reinforcements? Anything from being kicked all night by your *in utero* companion after you selfishly insisted on that extra spicy burrito to being hugged and kissed a lot and even to seeing fleeting evidence of emotional intelligence in your own kids. Just a few days after the dessert debacle, I was sitting at the kitchen table with Joey, now a calmly captive audience, as I complained about the nasty telephone treatment I'd just gotten from a cranky source on a magazine story. Joey listened and then coolly suggested, in what I could only hope was an echo of my own recent performance: "Maybe he was having a worse day than you were."

Initially, you have few clues about how well you're learning. You leave the hospital carrying a dependent lump who can't even make eye contact, except with your breasts, much less express his needs beyond a cry. It's up to you to figure out whether the tone, pitch, and volume of that cry means he's hungry, wet, tired, or bored, or perhaps in pain from some rare condition you really should have been reading up on. Or sometimes there's no right answer to the quiz, and all you can do is stay up into the wee hours, singing, rocking, feeding, and changing diapers, until one or both of you surrenders to exhaustion.

It's all part of emergency training in your particular infant's quirks and cues that can make for more intelligent parenting down the line. Your new sensitivity may eventually give you insight into other people's cues. But you will never study anyone quite so closely: At two weeks after delivery, mothers on average report spending nearly fourteen hours a day focused exclusively on their infants. (Fathers spend roughly half that time.) "You've really got to be paying attention to the littlest of movements—eye movements, changes in muscle tone," says Barry Lester, a pediatric psychology professor who studies infant development at Brown University. "You have some babies crying, and you pick them up in one position and they like it, in another and they don't. It all teaches you to be more sensitive to behavior in general."

Gradually, just as your arms grow stronger with repeatedly picking up a baby who weighs more each day, your brain is given a graduated workout as your kids pitch you ever more skillful curveballs. You grow along with them, responding to their changing needs as you call on different capacities in yourself. Once your child starts talking, for instance, you may find yourself adopting the empathetic style of speech that scientists call *motherese.* Parents automatically raise their voices about an octave in pitch; as they do so, they slow their delivery and use a simpler vocabulary and more repetitions, which they tailor to their children's ability to understand. Not only do babies respond more happily and attentively to this style of language, but it seems to be just what they need to learn human speech efficiently. One study, involving six-month-old babies in Japan, found that deaf and hearing babies alike responded better even to sign language when the gestures were made in an exaggerated kind of motherese. In contrast, a test of depressed mothers found they failed to modify their speech when addressing their babies, leading the authors to speculate that insufficient motherese might help explain how depressed mothers sometimes pass on emotional problems to their children.

Of course, almost as soon as children can form sentences, they become their mothers' faithful monitors, constantly calling attention to weak points and character flaws, sometimes by protest, sometimes by embarrassing imitation, and almost always with that potently reinforcing combo of punishment and love.

Anne Lamott has described this dynamic in a striking passage about conflict with her toddler son Sam:

> The fear is the worst part, the fear about who you secretly think you are, the fear you see in your child's eyes. But underneath the fear I keep finding resiliency, forgiveness, even grace. The third time Sam called for me the other night, and I finally blew up in the living room. . . . I lay on the couch with my hands over my face . . . and after a minute Sam sidled out into the living room because he still needed to see me, he needed to snuggle . . . like the baby spider pushing in through the furry black legs of the mother tarantula, knowing she's in there somewhere.

Empathy: The Root of All Mothering

Dan Batson, a psychology professor who specializes in the study of empathy at the University of Kansas, calls it an "other-oriented emotional response elicited by and congruent with the perceived welfare of another." In other words, empathy is different, and more altruistic, than the mere sharing of feeling, which is more appropriately labeled "emotional contagion." In this way, it is also a more conscious, purposeful emotion, which can be a foundation of trust and cooperation, and which is thus rightfully seen by Yale's Salovey and others as a component of emotional intelligence.

While it may seem a quintessentially human trait, empathy can actually be found in many mammal species. When researchers used a harness to suspend an albino rat in the air, another rat, watching the first one flailing in distress, pressed a bar to lower him back to safety. In a separate experiment, a rhesus monkey learned to pull a chain to be rewarded with food, but then saw his action was delivering an electric shock to another monkey, whose face he could see. The first monkey stopped pulling the chain for several days, preferring to go hungry than to cause another's pain.

To be sure, not all mothers feel empathy, and very few mothers feel it all the time. An image comes to mind of a scowling woman on skates I once watched: She strapped her screaming toddler into one of those precarious carriages that can be attached to bikes, but which in this case was attached to her waist. As she took off, the carriage wobbling, her hysterical daughter shrieked, "No, Mommy!! Not the roller blades!"

For most of us, however, empathy frequently informs our earliest days with our infants as we try to figure out what they need, how to comfort and satisfy them. As we practice this emotion again and again, it influences our brains—perhaps explaining why, in the Swiss brain-scan study described in Chapter 3, parents' brains became more active than those of nonparents at the sound of a baby crying. "The brain changes when you adopt a new mode of behavior or begin operating from a different frame of logic, or become engaged in highly emotionally charged learning," maintains Michael Merzenich. "All generate changes in brains that grow in strength as the behavior is practiced."

Where might a brain-scanner actually locate those changes? So far, scientists have only theories, based on the rudiments of what is understood about the brain's emotional equipment. Imagine, for instance, what might be happening when a baby smiles, and his mother smiles back. Parents share their children's moods all the time: wincing along when a son gets his measles vaccination, grinning when a daughter opens her birthday gift.

Kelly Lambert, the Virginia neuroscientist, recalls a time when she was changing her daughter Lara's diaper:

> She was on the changing table, so our faces were pretty much locked as I was talking to her while doing the changing part. All of a sudden she got some horrible gas pain or something and had this full blown crimson-red crybaby face. Without any conscious processing that I was aware of, I immediately found myself making the same crybaby face—and I remember saying to myself, wait a minute, why am I doing this? I felt like I was on empathy overdrive.

Lambert's empathy equipment really may have been moving into high gear. It has been known for some time that if a person sets his face into a certain expression, he will feel the emotion that fits it. Smiling has been shown to make people feel happier. And imitating another's expression, consciously or unconsciously, is a particularly potent way of "catching" an emotion, as scientists have recently been discovering.

Using brain scans, a team of University of California at Los Angeles neuroscientists tracked the responses of eleven subjects who looked at pictures depicting the facial expressions of six emotions: happiness, sadness, anger, surprise, disgust, and fear. Some of the subjects were asked to imitate the depicted expressions; others were asked simply to observe. The researchers found that imitation as well as mere observation activated a mostly similar circuit in the brain involving areas of the cortex involved in motion and two important emotional brain centers: the insula and amygdala. But imitation resulted in a higher level of activity than mere observation.

Kelly Lambert's crybaby face, as she felt her daughter's pain, was just the type of reflexive expression that prompts the brain to heighten empathy, according to the study's leader, Marco Iacoboni, a professor of psychiatry. Iacoboni had previously found that the cortical areas that were activated in his study are important in imitation. They're connected to the limbic system via the insula, which Iacoboni believes may be the key relay station in translating imitation into emotion.

Iacoboni suspects this circuit can either be damaged or strengthened, thus affecting a person's ability to empathize. Specifically, practice in imitating and being imitated in return might help strengthen the neural circuits active in empathy, he says. In autistic children, who are notoriously lacking in this social skill, being imitated encourages more friendly behavior, Iacoboni notes. "They get closer to you, they touch you more," he says. In the same way, a mother constantly exchanging expressions with her child might also be reinforcing her capacity for understanding her own child's feelings, and eventually those of other people.

"I absolutely believe that if you're more empathetic with your children on a repetitive basis, you'll be more so with others as well," Iacoboni says. "The underlying neurophysiological mechanism that you are using is the same. . . . So these regions of the brain become more reactive to stimuli, meaning probably you'll imitate others more automatically. You simulate others' actions, you keep empathizing, and it becomes a virtuous circle." Together with other scientists studying "positive" emotions, Iacoboni believes that a fundamental part of the brain's empathy equipment relies on recently discovered specialized nerve cells in the motor cortex. These so-called mirror neurons fire no matter whether you're doing something yourself or just watching someone else in motion: It's a "monkey see, monkey feel" dynamic as both of you exercise similar parts of your brain. Two conditions appear to make this reaction more intense: how emotionally close you are to the person you're watching, and whether you've both done or felt something similar in the past.

Tania Singer, a researcher at University College, London, has supervised *f*MRIs of sixteen women who, for both of these reasons, were expected to have a particularly strong empathetic reaction. All sixteen were scanned twice. On the first occasion, each woman received a painful electric shock to her right hand as she watched a computer screen that warned her when the pain would come and how intense it would be. Next, the women were scanned while their husbands received similar shocks. They could see the warning on the computer screen, but could not watch their husbands' faces.

Singer's scans showed that the same part of the women's brains lit up when the women expected to be hurt and when they expected their mates to suffer. When they described their experiences, they said they didn't actually feel the same pain, but did feel the same level of distress. As in Iacoboni's study, Singer saw activity in the insula, in addition to the anterior cingulate, which has been likened to a neural alarm system that draws the brain's attention to sudden distressing changes in the physical or emotional environment. This region lights up in response to pain—and it is also active when a mother hears her infant's cry.

Applied Emotional Intelligence

In the early 1970s, the noted pediatrician T. Berry Brazelton, then at Harvard Medical School, intensively studied mothers' interactions with their newborns. He wrote of one woman who was particularly skilled in reading her baby's cues and anticipating when he needed a break from the intensity of their communication:

> She relaxed back in her chair smiling softly, reducing other activity such as vocalizing and moving, waiting for him to return. When he did look back, she began slowly to add behavior on behavior, as if she were feeling out how much he could master. She also sensed his need to reciprocate. She vocalized then waited for his response. When she smiled, she waited until he smiled before she began to build up her own smiling again. Her moving in close to him was paced sensitively to coincide with his body cycling, and if he became excited or jerky in his movement she subsided back into her chair. . . . she seemed to be teaching him how to expand his own ability to attend to stimulation . . . [and] gave him the experience in pacing himself in order to attend to the environment.

An estimated 90 percent of all human communication is broadcast without words—in the lift of the eyebrows, the pursing of lips, the sunny or sarcastic inflection of a greeting, the bounciness in someone's step. For most of the first two years of life, children rely almost completely upon expressions, gestures, and nonsense words. Responsible parents have to become expert decoders.

Another Harvard study in the 1970s offered one of the earliest hints about how mothers themselves might gain from such interactions and so become more emotionally intuitive with adults as well as with babies. Robert Rosenthal, a research psychologist, had developed a test called the Profile of Nonverbal Sensitivity. He showed volunteers a series of videos depicting a young woman expressing a range of strong emotions. Alternately, she seemed hateful, jealous, apologetic, or maternal. But there were no clues from words; the videos were silent.

In one phase of the study, Rosenthal and his collaborators gave the test to eight mothers with "pre-linguistic" toddlers and six who were married but childless. They found that the mothers had a distinct advantage, which was replicated in a similar test six months later. The researchers cautioned that they couldn't say without a doubt that it was motherhood itself that had made these women more emotionally literate. There was always the possibility, they acknowledged, that more emotionally literate women are more likely to become mothers. Nonetheless, Rosenthal and his colleagues ventured a conclusion: "Infants appear to 'educate' their parents nonverbally," they wrote.

Many mothers I interviewed agree with this notion, several offering examples to illustrate their sense of having gained skill in reading nonverbal clues. Lori Willis, introduced in Chapter 7, accurately predicted that her eighteen-month-old son, Alec, would need medical care, for instance, just before he had to be hospitalized for gastroenteritis. "A week before he got sick, I could tell something was wrong," she said. "He wasn't being lethargic. There was just something with his eyes, a dullness there. He has these bright baby blue eyes, and the whites were just dull-looking."

And in San Francisco, Susan Kostal, a journalist and mother of three, told me how she had applied the nonverbal literacy she had learned with her children to an adult neighbor in distress: "A woman who seemed pretty lonely lived across the street from us, and from time to time she'd call to ask me to do favors," Kostal said. "One afternoon, she called and asked if I'd pick up her mother at the bus station, but the way she asked was different from before. Her tone of voice had changed; it was more like a command than a request." Slightly annoyed, Kostal nonetheless drove to the station. On her return, she saw police cars parked outside her neighbor's home. "I figured out right away that she had tried to kill herself," she recalled. "I went straight over and asked the police, 'Did she do it?' The cops were very intrigued about just how I'd known that."

Kostal's explanation was that she was simply more tuned in, and had learned to be so from her young daughters. "When a person's behavior seems out of place," she said, "that usually means they're getting sick, or about to make some big developmental leap. I didn't care that much about noticing these things before I had kids. But when you're a working parent, whether or not someone's getting sick is going to govern your life for the next forty-eight hours."

This extra vulnerability to what might be going on with someone else could be why many mothers may also develop expertise in a skill closely related to empathy, but a bit more cold-blooded. This is Theory of Mind: the art of figuring out what someone inherently different from you is thinking. Let's put it this way: It is said that when two animals stare into each others' eyes for more than ten seconds straight, they're preparing either to fight or to make love. To this must be added a third possibility, one in which an animal is trying to figure out just what its offspring was thinking when it drew on the wall, forgot to eat lunch, or got those weird tattoos. Like empathy, Theory of Mind involves reading another person's feelings from mostly nonverbal cues: a glint in her eyes, or the way he keeps his hands in his pockets. But whereas empathy implies an emotional intention to help, Theory of Mind is more of a removed, cognitive approach.

Paul Ekman, an author and a psychologist at the University of California at San Francisco, is famous for producing a remarkable tool for would-be Theory of Mind experts. He has documented thousands of the tiniest facial expressions that give clues to what people are thinking and feeling. The ability to read these fleeting "micro-expressions"—lasting as little as a quarter of a second and involving varied combinations of a total of forty-three facial muscles—is fundamentally important when judging another person's sincerity. Ekman has not looked at whether parents are any better than non-parents in this department, but maintains that, as with most emotional skills, you can improve your literacy with motivated practice.

Something helpful for a mother to keep in mind as she exercises her Theory of Mind is that her child may not only be figuratively on Mars (if he's male) but in another neurological galaxy, because he's not yet an adult. "I always remind myself that they're not fully myelin-ated yet," my psychiatrist brother Jim once said, while observing his five-year-old go ballistic over the loss of a plastic Transformer robot. Jim was referring to the brain's "white matter": fatty myelin sheaths, similar to insulation around wires, which encase axons, projections of the neurons that transmit information across synapses. In childhood, this insulation is quite thin, resulting in speedier neural connections. This may help explain kids' greater impulsiveness and more risky behaviors, especially when you also consider that the prefrontal cortex, that evolutionarily newest part of the human brain in charge of setting priorities, and, again, of subduing impulses, is the last to mature. By the time that brain region is fully developed, your child may be in his twenties, and you will have found many opportunities for practice in three other emotionally helpful skills.

Self-Restraint, Conflict Resolution, and Spin Control

In the daily practice of parenting, as you urge, nag, and nudge your child along the narrow path of civilization, you'll most likely learn three specific techniques you can employ to good effect with other people.

The first is self-restraint. Although this is one of parenting's hardest tasks, there will be plenty opportunities for practice, if you have a normal,

that is, conflict-prone, relationship with your child. Mothers of two- to three-year-olds have been shown to voice a command or express disapproval on average every ninety seconds, and conflicts between parents and young children have been reported at an average rate of 1.5 to 3.5 times an hour. (Both these rates strike me as rather low, assuming the parents and children are actually *in the same room.*)

A mother's capacity for self-restraint will be tested repeatedly on these occasions. Unless you have model children, you may find yourself frequently wanting to yell, swear, hit, swerve off the road, or fly to Rio, all impulses that the better part of you knows you must control. You might console yourself with the knowledge that, in the process, you're exercising your brain's frontal lobes, the part of the cortex directly behind the forehead and extending to the crown of the head and toward the ears.

Scientists made a great leap forward in understanding the role of the frontal lobes in the mid-nineteenth century, thanks to an unlucky young railroad worker named Phineas Gage. In September 1848, Gage survived a horrible accident in which an explosion shot a steel tamping rod (3.5 feet long and weighing 13.5 pounds) through the side of his head, seriously injuring his left frontal lobe. He recovered sufficiently to return to work some months later, but his bosses soon saw how much he had changed. "Gage was no longer Gage," as his friends put it. Previously courteous, efficient, and disciplined, he was now impatient, fitful, and profane, unable to put the brakes to his impulses. His doctor concluded: "The equilibrium between his intellectual faculties and animal propensities seems to have been destroyed."

Although the frontal lobes are some of the hardest working parts of the brain, even in people who aren't parents of hot-tempered kids, skillful self-restraint probably also recruits other areas. Buddhists, for instance, talk of the art of "recognizing the spark before the flame" when learning to discipline their emotions—an art that presumably involves more sophisticated mechanics than merely braking one's impulses to strangle or rant.

Assuming however, that your frontal lobes temporarily fail, you may have the chance to practice a second and more refined emotionally intelligent technique: the capacity to resolve conflicts and smooth ruffled

feathers. At Harvard Medical School, Edward Tronick, a developmental psychologist, has performed experiments with mothers and infants in which mothers are asked to sit and face their babies, but suddenly not to respond at all. Infants, quite naturally, react strongly to this "still-face" behavior. They may raise their eyebrows, turn away, face the mother again, smile, or fake a cough. But eventually, many collapse into sadness. The mother must then come back and interact, trying to make things right again. Depending usually on how the relationship has progressed until then, some babies will remain upset, holding back forgiveness, but some readily come around at the mother's prompting.

"It's a great lesson for adults," says Lester, who has closely watched mothers and babies interacting in the clinic he runs for infants with colic. As most of us know all too well, few, if any, relationships are free of conflict. Thus, Lester notes, "what makes for a successful relationship is *repair*, and it's obvious how you can translate that to adults. If you can do it with your baby, you can do it with your partner."

In times of conflict and of peace, and especially in political campaigns, a third emotionally intelligent skill—reappraisal, or spin-control—can be of tremendous use. A kind of learned optimism, reappraisal is "changing the way we think to change the way we feel," says Kevin Ochsner, a neuroscientist at Columbia University who has studied its underlying brain dynamics.

In managing your own emotions, a prerequisite to coping with the emotions of others, the ability to look on the bright side is key. Joan Didion, the erudite observer of American culture, long ago wrote that "we tell ourselves stories in order to live." The mother's narrative might run like this: "Ouch, that hurts!" "My, he's strong!" "I'd like to kill him!" "Then again, he's my son." "He's four years old and still not potty-trained!" "He is an individualist who won't be cowed by authority." "Why can't he stop arguing?" "He may become a rich and successful lawyer!" "Motherhood stinks!" "Perhaps motherhood is improving my brain."

One of my favorite parental spin-control stories comes from my brother Jim, not normally an optimist, but an extraordinarily loving father. One evening he came upon his six-year-old son in the bathroom,

sobbing as he urinated on the wall next to the toilet. Jim quickly realized that the boy had accidentally dropped his Yu-gi-oh! card into the toilet and had to make an "executive decision," as my brother put it, about what to do next. "He is such a thinker!" Jim admiringly concluded.

Ochsner and his colleagues have been studying people's facility with reappraisal by brain-scanning volunteers as they look at initially disturbing photos while hearing verbal explanations that cast the photos in a cheerier light. One picture, for instance, shows a group of women weeping outside a church, a scene that might easily give the initial impression of a funeral. But the volunteers are told that a wedding has just taken place, and the women are really weeping with joy. Based on the results of these experiments, Ochsner believes that the prefrontal cortex, the anterior or front part of the frontal lobe, underlies this emotional-management function and actually dampens activity in the excitable amygdala when people try to put a positive spin on a bad situation.

How might this way of thinking—including even my besotted brother's thinking—be "smart"? Mainly because optimism is a great motivator, making it a key part of emotional intelligence. The psychologist and best-selling author Martin Seligman, who has written extensively about optimism, and in particular "learned optimism," has found that optimistic insurance salesmen sold 37 percent more insurance in their first two years on the job than did pessimists. Optimism keeps parents plugging away to care for and nurture their children, trusting that there will be at least some emotional dividends one day. It also helps people bounce back from rejection and defeat.

Studies in past years have shown that depressed people have heightened activity in the amygdala and decreased activity in the prefrontal cortex, Ochsner notes, leading him to wonder whether they're constitutionally less able to tap into networks for reappraisal. Could you stimulate these networks with drugs to create a more optimistic worldview, he wonders? Or could you simply do it with lots of practice, the kind in which motivated mothers engage? Science has yet to deliver clear answers to either of these questions. Yet there is widespread agreement with Davidson and Ekman's argument that people can improve emotional skills by making a conscious effort. Ekman, who, like Davidson, has worked with Buddhist monks, reverently refers to the Dalai Lama as a "Mozart of the mind." He adds: "We can't all be Mozart, but we can be musicians."

Evolving Emotional Intelligence

As human mothers throughout history have nurtured their offspring and taught them to get along with others, thus increasing the chances they'll give us grandchildren one day, something encouraging has happened to our species: We've become, across the centuries, more cognitively complex, empathetic, and optimistic.

Paul MacLean, the NIH researcher, has argued that mothers' behavior has guided the social evolution of animals up from the lizards, who didn't even stick around long enough to warm their eggs, to today's human mother, who starts worrying about college tuition when she firsts hears her baby's heartbeat. Although the strategy of those at the bottom of the evolutionary chain is still to maximize the number of fertilized eggs, human mothers invest their time and energy in relatively few embryos, whose bigger brains and longer childhoods serve to prepare them to navigate their increasingly complex social environment.

The human baby enters the world with a brain that is just one quarter of its future adult size, a circumstance that allows his head to pass safely through his mother's pelvis, yet leaves him utterly dependent at birth. The rest of that brain's development will take place outside the womb; most babies can count on years of being toted around, stuck in high chairs and car seats, and extensively supervised and trained. This compact is at the core of our progressively more caring culture.

The evolutionary leap that made this possible occurred on the day when some reptile's daughter's chest secreted the drop of liquid that was the precursor to a mother's milk, explains Sarah Hrdy. That marked the end of the every-reptile-for-herself school of parenting. "There was nothing about the origin of lactation to indicate that it would ever have anything to do with the evolution of intelligence," Hrdy writes, "but it did." Nursing gradually obliged mothers to stick around with their kids, creating an intimate social relationship absolutely unlike any that had come before it. The long periods of closeness, as Hrdy writes, provided "the chance and necessity for 'social intelligence' to evolve." And, throughout most of human history, social intelligence, though not the only intelligence, was by far the most important kind.

This change didn't just affect babies; it hugely, if gradually, increased the demands on mothers. On top of nine months of pregnancy, women eventually were investing years in nursing, feeding, and protecting dependent children, escalating responsibilities that probably led to an "intellectual arms race around parenting," as Cort Pedersen, a professor of psychiatry at the University of North Carolina at Chapel Hill and a pioneering expert on maternal behavior, theorizes. "To be a parent meant that you now had to do more than give birth, but find food for your offspring and protect them from dangers in the environment," he says. "That's obviously much more complicated than just looking out for oneself. So given that view, it's not surprising that becoming a parent would notch up mental ability."

According to one path-breaking line of research, mothers may steer the evolution of intelligence and social behavior not only by their caring behavior but with their very genes. Experimenting on specially bred mice, Barry Keverne, at Cambridge University, has shown that genes inherited from the mother have more influence on the newer "executive" portions of the brain—the parts that are key for social interactions—than do those from the father. The mice blocked out for maternal genetic influence bore babies with bigger bodies and smaller brains. Those blocked out for paternal influence wound up with the reverse: babies with big brains and smaller bodies.

Evolution, of course, creeps forward over eons, not hours. Yet in the realm of each parent's life, there are chances for great leaps forward as we seek to correct for our own parents' mistakes and learn from our children's needs. In teaching them, we do the more difficult work of teaching ourselves. And the potential benefits of all this learning aren't by any means limited to biological mothers.

Part Three

SO WHAT?

CHAPTER
9

Mr. Moms and Other Altruists

Paybacks to Proxies

[A] good part of what gives motherhood its power . . . is not the selfless love of another, but . . . an opportunity to know and love a transformed self. . . . Men also have this opportunity, for there is nothing more inclusive than the idea of transformation.

GORDON CHURCHWELL,
IN *EXPECTING: ONE MAN'S UNCENSORED MEMOIR OF PREGNANCY*

MINUTES AFTER THEIR babies were born, a group of twenty brand new fathers held their infants against their naked chests. Twenty other fathers stood by and merely watched their wives embrace their babies. Two weeks later, the dads, who were taking part in a controlled experiment at the Karolinska Institute in Stockholm, were asked to talk about their new children. "The difference was very impressive," says Kerstin Uvnas-Moberg, the Swedish neuroendocrinologist introduced in Chapter 3, who led the experiment. "The ones who'd held the babies skin-to-skin would say things like, 'My baby is the best in the world, absolutely fabulous.' The others were also positive, but they didn't use these dramatic words. They would say things like, 'He's a nice boy and I hope he does well. The words they used clearly had different emotional value.'"

Even as research on the maternal brain remains avant-garde, scientists have already started to probe how fathers and "alloparents"—related or unrelated caregivers—can be changed, down to their biochemistry, by

oriented to others." Part of this process, he believes, is that parenting obligates adults to articulate and explain their own values, after which acting on those values becomes more natural. "That's why you see many adults going back to church when they have kids," he says.

Scott Coltrane, a sociologist and fatherhood expert, notes that reality has long surpassed the gee-whiz Hollywood presentations of Mr. Mom. In contemporary life, millions of devoted fathers are spending more time taking care of their own children, and being personally transformed by the experience. As a group, human fathers are models of paternal care, compared to roughly 90 percent of the mammal world, in which the standard is for dads to say "Ciao!" after conception. The human standard has long been much higher than that and is growing even higher, though many fathers, in practice, still take the "ciao!" route. In the process, human culture is also being changed, even if that change is coming much too slowly for many women still bearing the brunt of the infamous "second shift," in which they work as hard as men outside the home while also shouldering the brunt of the domestic load.

"Two things are happening at the same time," says Coltrane, an associate professor at the University of California, Riverside, and author of *Family Man: Fatherhood, Housework, and Gender Equity.* "There are fewer actual dads, but the dads who are in the picture are stepping up to the plate more." In other words, since the full-scale entry of women into the marketplace, marriage rates have fallen, divorce rates have risen, and consequently the average man is spending less time living with children than before. Yet in the families that do stay together, the average father is spending more time interacting with his children than at any time since researchers began collecting data. On average in the 1960s to the early 1980s, Coltrane says, fathers interacted with their kids about a third as much as mothers. By the late 1990s, they were interacting with their kids two-thirds as often as mothers on weekdays and more than four-fifths as much on weekends. Census figures show that one in four fathers is now caring for his preschooler while the mother is working. Meanwhile, more than 2 million children are reared primarily by their fathers.

Scores of these hands-on fathers have told Coltrane how their children have changed them in positive ways. Compared to their own fathers, they say they are more likely to hug their kids and tell them they love them. And many say they've found themselves working on personal problems, primarily anger management, in time spent with their children. Coltrane is convinced that men feel "safer" revealing vulnerabilities with young children, whose affection may seem more unconditional than that of their spouses. "Most men are reluctant to admit that they are deficient in any way," he says, "but according to the fathers, family work literally forced them to deal with various personal and relationship issues, and most welcomed the opportunity." A mail clerk whom Coltrane interviewed, for instance, told how his children helped him acknowledge and overcome his shyness. "I see that in my kids and it helps me work on it, too," he said. "They say, Daddy, would you come with me to go up to Sarah's door?' And I feel embarrassed . . . too, but I'm trying to get my kid to understand how to deal with people, so I do it."

If research on mice has any bearing on humans, modern, engaged dads may be gaining some of the same learning and memory advantages from parenthood as have been found in maternal rats. The key appears to be the degree of involvement with the children. In 2004, Kelly Lambert, the researcher at Randolph Macon College in Virginia, and her students compared the biparental California mouse *(P. Californicus)* with the deadbeat dad deer mouse *(P. Maniculatus)* and found that after each was exposed to an alien pup, the California mouse displayed comparatively more efficient foraging skills and less fearfulness, as he also more readily took over the care of the youngster. Lambert has found some similar improvements in biparental marmosets. The fathers, who tend to be especially domestic—the females always bear twins, making paternal babysitting a must—did better than bachelors in remembering the location of hidden Froot Loops.

In rats, in which mothers do the bulk of childcare, foster fathers experience a burst of new brain cell growth after being exposed for several days to newborn pups. This is similar to what happens to rat foster moms, except that the foster mothers outpace the foster fathers in cell growth by a ratio of more than 16 to 1. "The dads seem to hit a glass ceiling," Lambert's colleague, Kinsley, wryly notes.

After hearing about these smart mammal dads, I looked back on my notes of interviews with particularly involved human fathers, and noticed that some had already told me stories about becoming more efficient and less anxious. Gary Harrington, a part-time lawyer and screenwriter who works from home as the principle caregiver for his two kids while his wife works full-time at her San Francisco law office, multitasks as much as any mother I know. On the phone with a client one day, he caught sight of his then four-year-old son out of the corner of his eye, and, as he relates, "It was like, yadayadayada, *don't lick that!*"

Clinton Lewis, a semiretired software consultant who is the main caregiver for his three sons, reports that he has mastered navigational feats such as wheeling two children around in two separate shopping carts while buying groceries. He also notes that he has become more attentive to details such as socks strewn on the floor. "Such items were beneath notice at another time in my life," he says, "but since I've taken on many household duties, I now notice when the kids leave clothes on the floor, and I'm hoping to sensitize them to this issue rather than letting them annoy others for decades before they discover the cause."

Yet another extraordinarily engaged dad, Kent Salisbury, an electrician who, with his wife, software designer Carole Gifford, has equally shared care of their son Sam, tells how Sam, as a toddler, trained his mental focus. "I was really surprised by the fact that I can do very physical work—bending pipes, yanking hammers—but I found it way harder and more debilitating to hang out with a kid for four hours," he says. "You have to be alert and attentive, every minute. You can't look away. He cries and you jump."

There were also days when Sam wouldn't stop crying, when Salisbury had to walk him around in circles for an hour or more at a time. "I'd sing 'Cast Your Fate to the Wind' for an hour," he recalls. "It wouldn't work, and that was my best song." Yet Salisbury believes the patience he learned through those episodes serves him well today. "I find I'm a lot less anxious than I used to be to jump in and say something offensive," he says. "I can sit back and shut up, not just with my wife but in a work situation. For instance, I often have to do free estimates, and there are times when I tell people I would really appreciate if you would figure out what you want me to do before I get there. Then I get there, and there is a husband and wife. He says: 'We need a receptacle right there for my TV.' And she says: 'But honey, I thought we said we were going to put it two feet to the right.' And I'm sitting twiddling my thumbs. It used to be that I'd get disgusted with people in that situation, and just say, 'Have a nice life!' and leave. Now I sit back and tell myself: 'This will get done.' Being with Sam taught me that."

That not all men find such fulfillment in caring for children is unfortunately clear from the numbers who leave families behind and from some recent extraordinarily sour fathers' memoirs. An Amazon.com reviewer of the 2003 book *Poor Daddy: Adventures of a Stay At Home Father* by Jim Hagarty, noted: "The only solid piece of advice he has to offer is that you must end this ridiculous experiment as soon as possible and do as he did: Reverse roles with your children's mother and get the heck back out into the rat race where you most definitely belong." Still, as Uvnas-Moberg's skin-to-skin study with fathers and babies suggests, early exposure to intensive fathering may encourage more of the same—perhaps for some of the same biochemical reasons as it does with mothers.

The Chemistry of Care

Anecdotes about male morning sickness and sympathy weight gain—known as the Couvade syndrome, from the French *couver* (to hatch)—go back many years. Yet it was only in 2000, thanks to groundbreaking research by two Canadian scientists, when we learned that many men may in fact be sharing some of the biochemical havoc of pregnancy. These fathers-to-be are under the influence of some of the same hormones, if not to the same degree. It's apparently Nature's way of getting men in a fatherly frame of mind.

Blood levels of prolactin—that "parenting hormone" associated with breastfeeding and reduced anxiety—surge in expectant fathers, as do those of estrogen, the well-known "female" hormone, according to findings by Anne Storey, a psychologist at Memorial University in Newfoundland, and Katherine Wynne-Edwards, a biologist at Queens University in Ontario. There is also a doubling in the levels of cortisol, known as the fight-or-flight hormone, but which Wynne-Edwards has noted might equally be called the "heads-up-eyes-forward-something-really-important-is-happening" hormone. Cortisol thus might help explain the revelation that brain-scanner Mark George experienced the first time he walked through an airport after his wife became pregnant. "It was amazing," he says. "I'm always in airports, but I'd never noticed before how many *kids* were there. Overnight, just walking from gate to gate, I was thinking, where do all these kids come from?"

Somehow, some of the profound chemical changes taking place in the mother-to-be had spread to her partner—but how? The most likely answer is pheromones, those subtle biochemical messengers that get their point across most effectively with people who are physically close over time. The more closeness, the more likely a man will undergo hormonal changes, bringing on such pregnancy symptoms as nausea and weight gain. Remarkably, nine out of ten men, in two studies, experienced at least one such symptom. Storey's hypothesis is that fathers who don't live with their wives wouldn't undergo similar changes. "There are probably several ways that we become good parents," she says, "but maybe this process of being together during the pregnancy just helps people along a little bit."

This fatherly drug trip continues and in some ways intensifies once the new baby leaves the womb. At that point, the levels of a man's testosterone, that intrinsically male hormone linked with competition and sexual and physical aggression, plummet by as much as one-third. Over the course of human history, this change probably helped calm men down and made them less likely to stray at the time their mates most need them. And if you define "smart" as a mindset that helps you and your children survive, this is certainly a way that fatherhood makes men smarter. High testosterone increases harmful cholesterol, raising a man's chances of heart disease and stroke. It also contributes to reckless behavior, such as during the "testosterone storm" that hits at the onset of male puberty, and during which men are four to five times more likely to die than women, particularly from car accidents, homicide, and suicide.

The kinder, gentler hormone levels of fatherhood can fluctuate, just as they do with mothers, according to a father's contact with his infant. A man's testosterone levels fall even further and prolactin levels shoot up, for instance, when he holds a crying newborn, or even a crying doll. Oxytocin may also be involved, even though, as noted in Chapter 6, this hormone's influence appears to be stronger in females. Uvnas-Moberg suspects that oxytocin made the difference in her study of fathers who embraced their babies skin-to-skin.

Comparing human fathers and nonfathers in 2002, Alison Fleming, the Toronto neuroscientist, found that the fathers responded more intensively, their hearts beating faster and hormonal levels changing, when they heard infant cries. Compared to the childless dads, the fathers also reported *feeling* more sympathetic and more alert. As it turned out, those who felt the most sympathy and need to respond had the lower testosterone levels.

One thing that may be happening with engaged fathers, though at this writing it's just a theory, is that the higher levels of prolactin in their blood may be creating a kind of virtuous cycle. As the father cares for his baby, his elevated prolactin increases the activity of neurotransmitters, such as those of dopamine and beta endorphin, enhancing a feeling of pleasure—and inspiring him to continue to give care.

Still another hormone at play when a man takes care of his young children is vasopressin, which in molecular structure is similar to oxytocin. We know this from studies of prairie voles, small furry rodents of exemplary domestic behavior. The males mate for life after their first sexual encounter, and subsequently stick around the nests, as caring, protective fathers. Immediately after their babies are born, vasopressin gene expression increases in male prairie voles—but not in meadow voles, a separate species in which the male is *almost* incorrigibly promiscuous. I say "almost" here because in 2004, researchers found that when they implant a gene that turns on receptors for vasopressin in the brains of meadow voles, these voles can be reprogrammed into faithful spouses and apparently loving fathers. The mind reels at the potential implications for the human war of the sexes.

A human father's circulating hormones gradually return to normal within a few months after his baby's birth. Yet depending on how closely involved he has been during the pregnancy, delivery, and postpartum havoc, he is surely a man transformed. Alison Fleming found one indication of this when she discovered that dads with two or more children have a greater increase in prolactin in their blood when caring for their infants compared to first-timers. In other words, parenting experience makes at least some permanent adjustments to a father's brain, just as it does to a mother's. Furthermore, the Harvard researchers mentioned in Chapter 8 who tested mothers in nonverbal literacy found that fathers, too, had made gains in this field of emotional intelligence, though their improvements were less dramatic. "If fathers begin to spend more time in child care, one might expect the . . . gains for the men to be more like those we have obtained for women," the researchers optimistically concluded.

Extending Families

"I didn't end up meeting Mr. Right, but I realized that didn't mean I had to give up the dream of having a child," says Laurie-Ann Barbour, who at forty-one adopted a four-month-old girl from Vietnam. Barbour's decision to "fulfill that something in myself that wants to love and give and nurture" is so common that, despite considerable legal and financial obstacles, some 125,000 children were adopted each year in the United States through the 1990s, about half of them by unrelated adults.

For Barbour, who had previously lived alone and held a variety of jobs, from office work to catering, having a daughter has involved becoming part of a more stable social network. The adoption inspired her to join a co-housing group in Sonoma County, California, where nearly fifty adults and twenty children form a community that shares meals, a garden, and childcare. "It's a wonderful feeling," she says. "If someone is crying outside, all these windows pop open, and people respond."

This scenario suggests that Barbour might be benefiting from higher levels of oxytocin, and surely more good-for-your-brain social interactions, all indirectly related to her choice to become a parent. And at least in her early days with her baby, she may also have experienced higher levels of prolactin. This "parenting hormone" has been found to rise in marmoset siblings that carry new infants, leading researchers to believe it has something to do with caregiving and skin-to-skin contact. And although at this writing there appeared to be no published research on prolactin in adoptive families, there's some reason to suspect that many an adoptive parent's contention that he or she is profoundly changed by contact with a new child has some biochemical basis. Luciano Felicio, a professor of veterinary studies at Sao Paulo University in Brazil, recounts that one week after he and his wife adopted a child, his wife's gynecologist asked to measure her prolactin, and indeed found it was higher than normal.

Hormonal influence aside, it does seems likely that adoptive parents, just like biological parents, may be mentally enriched by the experience of caring for children. Research on rats underscores the potential. As described in Chapter 5, "foster" rats made gains in memory and learning ability that were similar to those of biological moms, albeit not as pronounced. The extra stimulation that babies brought into their otherwise uneventful lives seemed to make the foster mothers more efficient. And as Craig Kinsley and Kelly Lambert found, exposure to pups also resulted in the sprouting of hundreds of new neurons in the foster moms' hippocampi. "The brain is stimulated; it's as if the mere presence of pups plugs into some parental circuit," says Lambert.

There is no denying, all the same, that the nine-month preparation of pregnancy and the experience of delivering a child give biological mothers an advantage in making what's usually a life-long commitment. In

1985, researchers who measured the security of an infant's attachment to its mother at fourteen months found no difference between children of matched groups of biological and adoptive moms. This, they concluded, suggested that a "hormonally primed" mother wasn't essential for *infant* development. Yet other research from the 1980s, this time focused on the mothers, found that adoptive mothers tend be more protective and to display higher levels of anxiety about parenthood than biological moms. Michael Numan at Boston College says this difference underscores the importance of hormonal changes that help reduce anxiety in biological mothers. Yet considering that adoptive parents are normally under relatively low life stress and enjoy strong social and financial support—allowing them to adopt in the first place—he speculates that they may usually adjust over time.

Susan Kostal, the San Francisco mother introduced in Chapter 8, fits this scenario. She is financially secure, and she has a good marriage and firm commitment to her Christian faith. All this made her feel well prepared, after giving birth to two daughters, to adopt a third, whom she met in an orphanage in the Ukraine, where Kostal's husband, Marlow, had been working. Lara, a pretty blonde toddler, was then already two and a half years old. She had been born prematurely, weighing just two pounds, was missing half of her right arm, and spoke neither Russian nor English. She seemed to know only one nonsense word, which she repeated over and over. Six months after first seeing Lara's picture, Kostal had found herself waking up in the middle of the night and wondering how the girl was. "It was unsettling," she says. "Nothing like that had ever happened to me before and I'm not a real rescuing type of person." Yet she and her husband both felt they had "been given a job to do," as she says.

Kostal knew she was taking on a lot, yet ended up surprised at the difficulty of the transition. She says she felt "no romantic connection or sense of destiny" upon meeting Lara. The toddler greeted her by throwing a spoon across the room at the orphanage, subsequently spit banana on her during the flight back home, and wouldn't sit on her lap unless bribed there by food. "After three or four months, I had compassion fatigue; my energies were waning, and all this effort was not being reciprocated in the way I'd learned it would be with biological children," Kostal says. "So while we knew what a huge issue attachment could be for a child, I stupidly did not consider the huge issues for Marlow and me . . . I would have liked to have had those early days with her to fill up all that good feeling and connection most mothers and infants have that carry you through the terrible twos and beyond. Without that, it really becomes pure behavior management."

Four years after the adoption, Kostal says she and Lara have grown closer, although the relationship remains much more awkward than the one she has with her other daughters. "In some ways, we're bonded because we've been to hell and back . . . and I know what makes her tick," Kostal says. "She has made a lot of progress, and while I know a lot is her own drive, I'm sure we've contributed. I don't think we've squelched her true self, and her true self is a wonderful, powerful entity."

Kostal is also keeping faith that the experience has been and will be good for her and her family. "Already, we're all more sensitive to each other and other people around us, having had this day-to-day experience of not everybody being like you," she says. "For me and Marlow, it's humbling. It has proven to us we don't know everything . . . I now understand whereas one kid can be told to settle down, another has to be told more emphatically—'Come over here; I need to talk with you,' with a hand on the back—so I'm less likely to make a snap judgment about a lot of other situations."

Fortunately, many other adoptive parents find the education that children can provide comes more easily. Tens of thousands of children are adopted each year, for instance, by adults who marry their divorced or widowed parents, and presumably have had the chance to understand fully what they are taking on. Although these new attachments can bring troubles of their own—jealousy being a major problem—the initially unsought addition of a child in someone's life can also deliver surprising benefits. "You taught me how to care until it hurts; you taught me how to smile again. You taught me that life isn't so serious and sometimes you just have to play," wrote U.S. Army Pfc. Jesse Givens, thirty-four, of Springfield, Missouri, to his six-year-old stepson, in a letter published after Givens's death in Iraq.

As many parents know, some of the most cheerful proponents of the benefits of caring for children are the unmarried aunts and uncles who can spend time without taking on the 24/7 responsibility. Single at forty-six, Lesley Koenig says she has no regrets about having kept an unwavering focus on her career, directing opera and ballet productions, in the years that her sister has become a full-time mother of three. Koenig refers to herself as the "super-aunt." "I've spent a lot of time with them since they were tiny, and they've learned to come to me with all the good stuff, the really fun questions," she says. "They talk to me about drugs and sex and all the things that would worry their mom out of her mind." Koenig is convinced that the experience has increased her own empathy and understanding. "I had always been so impatient with other people who just don't get it," she says. "But being with them over time has taught me just how different people can be. Now, instead of being so quick to judge, I don't assume that just because someone talks right that they get it."

An increasingly common subset of modern adoptions are grandparents who wind up being the sole providers for their grandkids, just when they thought they might start to enjoy their retirements. Some resent the imposition, but others find it a source of focus and meaning.

In San Francisco, Suzan Houseman ended up caring for five of her grandchildren, all younger than six years, instead of living the more leisurely life she had expected. This was an especially tall order, because Houseman suffers from a degenerative joint disease that makes walking painful, sleeps with an oxygen tank, and lives month-to-month on her disability payments. Two of her grandkids are autistic; two others have learning disabilities. But their parents proved unable to provide for them at all. As Houseman says, "It's a lot of work. I can't help but pass out at the end of the day. But being with them gives me a reason to get up every morning. I have to get up. I must get up. They need me."

Healthy Altruists: When Nice Guys Finish First

At a center for homeless people with AIDS, in a part of San Francisco where tourists rarely linger, Alyssa Nickell massages the neck of an aging biker. The biker's earthly possessions, contained in plastic garbage bags, surround his feet, and the stench of his life in the streets fills the room. Nickell is not wearing gloves, nor does she appear to be holding her breath. A thirty-two-year-old theology student, she has been doing this work with a group called Care Through Touch for six hours a week for the past three years. "I'm not sure I'd ever think of it as altruistic," she says, "because I'm so fed by it. There are certain days when I'll feel like, oh, I'm not sure if I have anything to give. But then I go, and the work leaves me feeling more awake and strong."

As Erik Erikson argued, you don't have to have children to engage in altruism, and, often, childless adults like Nickell have more time and energy to "mother" other *adults* in need. In doing so, they may share some of the healthy rewards of parenting—with the advantage that they can go home in the evening and watch sitcoms, if they want. The benefits they get can include a heightened sense of self-worth, stress relief, gains in empathy that come from practice, and even overall health.

Sharon Lamb decided early in life that she did not want to be a mother; her career as a nurse was taking off and she was giving talks throughout the country: "I knew that if I'd had kids, I'd have wanted to devote a lot of time to them," she says. Instead, she focused her energies on becoming a leader in her distinctly altruistic profession, and in addition volunteered as head of a San Francisco support group for patients suffering from brain tumors. She has counseled strangers by phone at night and on the weekends, and held the position of secretary of the National Brain Tumor Foundation. "Most of the people I deal with are bound to die soon," says Lamb, who turned sixty in 2004. "But I think that even if they're facing death, it is comforting to know there'll be someone there for them. And this reduces my own stress—it maybe makes it not so hard for me to deal with the thought of my own death. Because I know I'm doing all I can."

For decades, evolutionary theorists have pondered why people like Lamb do what they do, sustaining the surprisingly prevalent behavior that has helped human society thrive. Altruism initially seemed to contradict the Darwinist perspective of the struggle for individuals, and their offspring, to survive. But beginning in the 1960s, leaders of what has been called the second Darwinian revolution offered new ways to think about what initially seem to be wholly selfless acts.

The biologist William Hamilton developed the theory of "kin selection" or "inclusive fitness," arguing that helping our relatives ultimately helps our genes survive because we share some of our genes with siblings, cousins, aunts, and uncles. Then in the 1970s, another famous theorist, the Harvard biologist Robert Trivers, came up with an explanation addressing the apparent mystery of why people and animals often help others unrelated to them. He called this behavior "reciprocal altruism," best expressed in the common saying, "You scratch my back, I'll scratch yours." That's to say, people and animals trade altruistic acts over a period of time, during which each party presumably keeps track of the other's behavior. Trivers suggested this score-keeping forms the basis of friendships. Other theorists have further extended these ideas, proposing other ways that we do well by doing good. For instance, even if we are not granted a reciprocal favor from an altruistic act, we might increase our own prestige, and perhaps win spontaneous favors from people who feel confident we'll live up to our generous reputation.

As most of us already know, trust and cooperation *feel* good, a skillful ploy of nature that can sometimes turn a Scrooge into a Santa. Neuroscientists have only recently begun to zero in on why this is so. Remember Paul Zak's finding, in a computerized game played by strangers, that oxytocin surges in people who display or receive signs of trust? Researchers who have scanned people's brains under similar circumstances have found that the mere sight of someone with whom they've just cooperated triggers increased activity in the brain's "reward" center.

At the same time, these warm, fuzzy, altruistic feelings may be physically good for you, as the Vaillant and Brown studies, among others, described earlier in the chapter, suggest. As part of the new focus on "positive" psychology, researchers increasingly over the past two decades have been turning their attention to the biological mechanics behind this effect. Spurring their interest is new funding, available from entities such as the Institute for Research on Unlimited Love—as well as the fact that, as baby boomers start to think about retirement, they're also devoting more thought to what kinds of activities support a longer and healthier as well as more meaningful life.

Zeroing in on the health benefits of altruism, researchers have noted that for many people, particularly the elderly, volunteering tends to increase their rate of physical activity, which in itself is good for the body and the brain. After that, there's something social scientists like to call "a sense of agency"—implying a feeling of control over one's environment. In one famous experiment examining this trait, elderly patients in a nursing home were randomly assigned whether or not to take care of a house plant. Those who tended the plant had a significantly lower death rate during the time of the study. Volunteering furthermore implies being social, which, as shown in Chapter 6, seems to help combat stress in ways that can also be good for your body and brain.

As she reached middle age, without a husband or children, Nancy McGirr deliberately chose a life that encompasses all these benefits. After spending the 1980s as a combat photographer, covering wars in Central America, she realized she felt isolated and sad, especially after having watched as two young photographers she had trained were shot and killed in El Salvador. The murders spurred her decision to move to a new country and "take a look at things that didn't have to do with war." McGirr settled in Guatemala, where she built herself a new career centered around positive emotions and children. In Guatemala City, she discovered thousands of scavengers living on the capital city's main rubbish dump and making their living by selling discarded plastic and metal. She began training the scavenger kids to use cameras, and, in 1991, founded Fotokids, a philanthropy that provides scholarships, classes, and cameras to children affected by war. Some of her students have since traveled throughout the world to exhibit their work. Many have been spared from lives of crime and drug addictions. "I'm really happy for these guys, and they've been part of my life now for twelve years, so you want the best for them, and that's what keeps me going," McGirr told an interviewer in 2003.

Like the brave mothers featured in Chapter 7, McGirr today is bold and ambitious on behalf of the children she has informally adopted. And like the mothers described in the chapter you'll read next, her children have inspired her to be supremely organized and creative at work.

CHAPTER
10

Better Than Business School
Mothers' Added Value at Work

If you can manage a group of small children, you can manage a group of bureaucrats. It's almost the same process.

A CORPORATE EXECUTIVE
QUOTED IN THE WELLESLEY COLLEGE
REPORT, "INSIDE WOMEN'S POWER"

BOOK EDITOR AMANDA Cook is one of millions of U.S. working mothers who use parenting skills on the job. A particularly useful tactic she picked up while caring for her toddler son, Aidan, relies on the seductive power of freedom of choice. "Rather than saying, 'No, you can't have the candy,' I ask Aiden, 'Would you like the blueberry or the strawberry yogurt?'" she explains. "The operative principle here is the 'illusion of control.'"

This same illusion works wonders in the office, Cook has found, where her writers now enjoy the option of delivering a manuscript in two months or four (but not six) or picking one of two subtitles (but not the title). It's merely one of many tools acquired in Cook's continuing at-home education: a rigorous course in the art of understanding other people, one of the most fundamental, albeit historically undervalued, kind of workplace smarts.

There are nearly 26 million working mothers in the United States today, roughly 40 percent of them with children younger than six. And to take liberties with that old song, "M" is for the million talents they bring to their jobs. Under the right circumstances, every one of the previously described attributes of a baby-boosted brain can be used to professional advantage. So can a myriad of more specific domestically acquired or enhanced skills, such as the ability to solve problems creatively, to exercise patience, and to recognize and nurture other people's best qualities. You might put it this way: All I Really Need to Know (at work), I Learned (by the time my child was) in Kindergarten.

Whether you're a CEO of a Fortune 500 company or a grocery checkout clerk, success at work and at home require proficiency in the same two basic skill sets—the logistical capacities that take you through the day with minimum bloodshed and maximum productivity, and the emotional intelligence that brings out the best from the people around you. This is truer than ever in light of profound recent changes in the United States that have led to what some economists call the "face-to-face economy." As blue-collar jobs transfer to lower-wage nations, as robots replace assembly line workers, and as much routine clerical work becomes computerized, employment is increasing in mainly two areas: service jobs and managerial/professional positions. These encompass tasks from waiting tables, to nursing the elderly, to trying cases in court, to running small businesses.

As mothers ever so slowly reach the top of many face-to-face occupations, they're becoming some of the most vocal advocates of their own crossover advantages. In San Francisco, for instance, Joanne Hayes-White, the first woman—and mother—to head a major city's fire department, believes that the practice of rearing her three children, all still younger than twelve, makes her smarter at work, adding that she has noticed the same extra-value in other mothers on her force. "There's a 24/7 aspect to this job that's very similar to being a mother," Hayes-White notes. "You have to learn to be very organized, and you have to be able to be flexible enough to respond on a dime. When that bell goes off, whatever you're doing, even if you're sleeping, you have to be ready to move. That's similar to mothering, too."

To be sure, we're still far from the day when a mother's child-rearing experience might be meaningfully acknowledged as an asset on the job. Although the famous "wage gap" between women and men has narrowed over the past several years—as a group, childless women earn about ninety cents to the dollar for men—women *who have children* earn much less: on average just seventy cents to the working man's dollar. This gap yawns even wider if you focus in on single mothers, or African American mothers, or mothers of more than one child.

The so-called Mommy Gap is only partly explained by measurable handicaps such as time taken out for maternity leave and other childcare requirements, says Diane Halpern, a psychologist and the director of the Berger Institute for Work, Family and Children at Claremont McKenna College. "If a woman stops out of the workforce for four years, her lifetime earnings will be much less than what they would have been if she had stayed in the workforce full-time minus four years of salary," Halpern says. "The penalty is greater than the time out."

What explains the extra hit? It may well be enduring prejudice, the unproven expectation that mothers won't be as loyal, as committed, or as competent as other workers. Princeton University researchers who conducted a recent opinion poll about working mothers discovered explicit evidence of this prejudice, concluding that "when working women become mothers, they unwittingly make a trade—perceived warmth for perceived competence. This trade unjustly costs them professional credibility and hinders their odds of being hired, promoted, and generally supported in the workplace."

This bias endures despite the paucity of research objectively measuring the productivity of working mothers versus other employees, or even whether mothers on the whole take more time off, a charge sometimes leveled by childless workers. Men seldom suffer the same prejudice; fatherhood usually *helps* their careers, giving them a cachet of stability. Mary Ann Mason, a law professor and the dean of graduate studies at the University of California at Berkeley, researched the influence of fatherhood versus motherhood in academic advancement. She found that for all fields of studies, married men with children were the most successful of any group at securing a tenure track position and eventually acquiring tenure. Married women with kids younger than six were the least successful.

There's still *some* good news for working moms in the overall economy: Gains in average pay have been greater for married women with children than for any other group in recent years, as this group has increased its numbers in the labor force more than any other since 1970 (barring a slight dip reported in a 2002 U.S. Census survey). Encouragingly, too, there are increasing signs that emotional intelligence, once viewed as a frill, is becoming more valued—and rewarded—on the job. "We've found that when workers have emotional intelligence, employers and coworkers see it as contributing to a positive environment, and it is correlated with important real-world outcomes—like getting a raise," says Peter Salovey, the Yale psychologist and emotional-intelligence expert.

Salovey and his colleagues collected direct evidence of this when they recently tested forty-four analysts and clerical employees from a Fortune 500 insurance firm on their emotional intelligence abilities. They found the highest scorers had already received larger merit raises, held higher company rank, and received higher peer and supervisor ratings than their counterparts.

Other recent studies have found similar results. The most successful U.S. Air Force recruiters, for instance, turned out to score highest on a test of emotional intelligence. Veteran partners in a multinational consulting firm who scored high on an emotional intelligence survey delivered $1.2 million more profit from their accounts than did other partners. In jobs of medium complexity, such as those of sales clerks or mechanics, top EQ performers were found to be 85 percent more productive than workers of average emotional intelligence. In more complex jobs, such as insurance sales or account managers, the difference rose to 127 percent.

As more employers recognize the value of emotional skills, they may also be slowly discarding old prejudices against working moms. "Companies are becoming less biased against them," says Cary Cherniss, a psychology professor at Rutgers, who counsels businesses interested in identifying and tapping their employees' emotional intelligence. "They're beginning to see that once people become parents, they really bring something back that is valuable."

What may also eventually help working mothers is increasing media attention on the crossover potential of their domestic talents. In recent years, the conversation has reached the point that serious attention has been given to the once-laughable idea of listing motherhood on a resume. Former Secretary of State Madeleine Albright, who took several years off working professionally to rear her three daughters, told one interviewer recently that she would "probably" put parenting on her resume were she looking for a job today. But let's take the idea a bit further. What subsets of skills that can lend value to most professions might a practiced mother claim? I suggest four stand out: an ability to coordinate a variety of tasks under pressure, dependability, leadership, and caregiving.

High-Wire Juggling

It's reasonable to assume that most mothers exercise many of the same brain structures while working at high-pressure jobs and dealing with their young children. Through engaged and often arduous, high-stakes training, mothers become proficient in meeting deadlines, coordinating multiple tasks, seeking creative solutions to impasses, dealing with frequent interruptions, and staying unruffled in a crisis.

Academic researchers increasingly are turning up evidence of a symbiotic relationship between engaged mothering and professional achievement. In 2003, a Wellesley College in-depth report on the experience of sixty exceptionally accomplished female professionals found that 20 percent of them cited mothering as a "training ground for leadership." Another 20 percent spoke of their own leadership style, or the style of others they respected, as being similar to mothering.

Of all the crossover skills shared by mothers, successful professionals and lesser-skilled workers, the one most often cited in the Wellesley report and in dozens of interviews for this book was time management. "As a mother, a firefighter, and a fire chief, you've just got to learn to juggle and prioritize," says Hayes-White, who, incidentally was interviewed by cell phone as she drove between appointments in her siren-equipped sedan. "I've really found the two jobs have a lot in common, mainly in that there is never enough time to do things."

Ann Moore, CEO of Time, Inc., told a radio interviewer that working mothers' efficiency makes them especially good employees. "I particularly love working mothers who have very young children at home, because they don't waste any time," she says. "They're the most kind of sleep-deprived, time-pressed people, and I find them to be really efficient. . . . I am always in awe of how much they can accomplish."

"I just don't have *time* to go back and forth and be neurotic about things," says Elizabeth Traut of the University of California at Berkeley. Traut is a biologist, a postdoctorate researcher, and a teacher who is also raising two spirited preteen boys: "Like, when I was a grad student, we all worried about what other people thought of us and that we were never going to know enough. As a mom, I know I'm *never* going to know enough, and that has made me much, much more focused and decisive."

One of Traut's tactics is to confine commitments, such as the lab class she teaches, to a specific number of hours each week, with no exceptions. She tells herself she'll do the best she can within that limit, and she says she's done just fine so far. Traut's husband, Jason, an elementary school teacher, says he's also seen a change in his wife, postmotherhood, in that she seems to read and "get" scientific literature more quickly. "When I know I only have a certain amount of time, it's like it eliminates the noise in the background," she says.

"Your intuition bumps up a lot," is the way Kathy Mayer, the Colorado Permanente doctor, puts it. "Like before, you had to ask question after question to find out what is going on with a patient, and now it's a lot clearer, sooner."

As a subset of time management, a mother's acquired skill in multitasking is also frequently cited by mothers themselves and by their bosses as providing added value on the job. Patty Ochoa, a grocery clerk and the mother of three, with two teens still at home, is rarely fazed when she's ringing up purchases—a job requiring her to memorize more than one hundred codes—while also bagging groceries, answering the phone, and taking care of other customers who are usually trying to catch her ear. "Things are flying at me from every side, but I'm so used to this that I'm on cruise control," she says with a laugh, adding, "and there's not a mother out there that doesn't know what I'm talking about."

That would include Catherine Gray, the president of an environmental think tank in San Francisco called The Natural Step. After several all-consuming years of work, she gave birth to her son, David Kai, at the age of thirty-six. Because The Natural Step was still in a "startup mode" and she wanted to give it her best, her plan had been to wait a couple more years. Yet, to her surprise, she says, she found that her new baby actually strengthened her executive abilities. On her maternity leave, she learned to combine cleaning her house and completing a conference call during the hour when Kai was down for his nap. "Then, when I got back to work, It didn't unsettle me anymore so much to have four or five balls in the air," Gray says. "Like, here's this one; there's that one; move this one. It all comes naturally."

Beyond time management and multitasking, motherhood can often help a woman acquire general competence at an impressive variety of tasks. This point is made almost every Mother's Day in a staple newspaper feature concerning what moms might be paid if society acknowledged the diverse jobs they do.

One of the most audacious estimates appears in the annual survey by Ric Edelman, chair of Edelman Financial Services, who has suggested fair wages for an active mother might run as high as $635,000 for a laundry list of jobs including childcare, cooking, housekeeping, pet maintenance, nursing, financial management, transportation, homework supervision, and . . . well, you get the idea. "Until I became a mother myself, I wasn't as engaged in the actual transferability of what you get," says Rayona Sharpnack, the Redwood City, California, women's leadership coach. "But when you have to give a keynote speech and lead a seminar and teach your daughter to ride a bike all in the same week, that's *building capacity.*"

Dependability: I Really Need That Job!

At the Reneson Hotel Group in Novato, California, nearly one-third of the 225 workers are mothers. Jennifer Wade-Yeo, who does the hiring, might have even more mothers working for her if the law didn't prohibit her from asking about family status. "The mothers are definitely more reliable, because they have that family to support," Wade-Yeo says. "The childless people can come and go. But the ones with kids, they need those benefits. They have to have their lives organized." Despite many employers' continuing perceptions that motherhood saps a woman's energy and commitment for work, the truth is quite often just the opposite. Many working mothers end up toiling all the harder—from financial need, gratitude, and sometimes just to take a break from hours picking up little Lego bits from the carpet.

At the San Francisco Fire Station, where 223 of the 1,600 uniformed officers are women—one of the nation's highest numbers—Chief Hayes-White says the twenty-four-hour shift work is noticeably harder on the nearly seventy mother fire-fighters. Yet she says they invariably make do, often trading shifts with other working moms, and that their overall devotion to their jobs "absolutely makes up" for the times when they do end up with scheduling conflicts.

Lori Willis, the divorced administrative assistant in Massachusetts, says she sometimes goes to work on Sundays, when her ex-husband has their son, just to burn off energy. "I do so much during the week, and I'm so used to going a mile a minute, that when I have the day off it's like, what am I going to do?" she says. "My brain just wants to do so much more." Other mothers agree that working at maximum capacity seems to increase their endurance and give them a more intensive level of functioning. Kelly Lambert, the Virginia neuroscientist, carried out her pathbreaking research on rats with Craig Kinsley while she was also writing a textbook, teaching, serving as chairman of her psychology department, and rearing her two young daughters. Yet, despite all this energetic output, she says she feels smarter, more productive, and less in need of sleep than at any other time in her life.

Especially for mothers lucky enough to have interesting jobs, going to work can be a refreshing respite from domestic duties, a contrast that can make them all the more effective in their professions. This may account for why some surprising recent research has shown that married workers, including those with children, who work shifts (i.e., outside the normal hours of 7:00 A.M. to 6:00 P.M.) have higher levels of life and job satisfaction than their unmarried coworkers. Sue Sommer, an emergency room physician who frequently works long and unusual hours, maintains that in her most demanding moments on the job, "with blood flowing and patients crumpling on the floor right in front of me," she can remain more composed than when things are wild and crazy at home with her two boys, aged six and eight. "There's a *system* in place there, and as a doctor, I actually have some control," she says. "I can ask someone to do something, and they do it."

The whole idea of even minimal control, for someone immersed in the chaos of rearing preteens, can be enchanting. "There's something about knowing that you see a patient, and that's done, and you see another patient, and that's done, and that when your shift is over it is all done for the day and you go home," Somers says. "In contrast, motherhood is a process where you rarely, if ever, see immediate results, and have to keep plugging on faith."

Arlie Hochschild, a sociology professor at the University of California at Berkeley, examined this phenomenon in her fascinating 2001 book *The Time Bind.* She found that many working parents at a company she studied were reluctant to reduce their workloads or take advantage of flextime or other programs offered to help them balance their lives—not just because they felt they needed the money but because work had become such a pleasant escape from the seemingly escalating demands at home. Her findings offer a worrisome picture of the state of parenting in the United States—a topic examined further in the next chapter—yet highlight how having kids can somewhat insidiously increase a worker's motivation on the job.

The paycheck is still work's main reward, however, and it is clearly a leading reason why many women are more serious about their jobs once they have children. Lori Willis says the birth of her son inspired her to try out for—and secure—a higher-paying position. Other mothers, confronting startling new financial demands from their children, acquire more training, take on extra shifts, and generally push themselves harder on the job.

Finally, there's another kind of responsibility—informing personal choices and character—that working mothers (as well as fathers) can acquire at home, and which often helps them on the job. In Ann Crittenden's book, *If You've Raised Kids, You Can Manage Anything,* Valerie Hudson, a political science professor at Brigham Young University, refers to it as "habits of integrity," suggesting you tend to pay closer attention to the moral weight of your every action once you realize you've got a highly impressionable and constant audience in your child. "Mothers have to craft a life that their kids can emulate," Hudson, a Mormon and the mother of six, told Crittenden. "They are fully visible to their kids. Sometimes I'll think maybe I can cut a corner here or there and do something an easier way, and then I'll think, no, they are fully aware of what I do, and I always have to be setting that example."

Leadership: The Motherly Art of Management

When asked which part of parenting helped most in her subsequent diplomatic career, former Secretary of State Albright answered, "Getting people to play well together!" Mothering, that is, provided the groundwork for her expertise in managing a huge bureaucracy and supervising highly emotional negotiations over foreign conflicts. In doing so, she would often be reminded of "children arguing and feeling that they can't understand the other person's side," Albright told *Mothering Magazine.* But she had learned, she went on, to get kids to stop their bickering and "try to figure out why the other person cares so much about that toy. So a lot of it is basic respect for the other person."

"Offices do for some instead of families, and for others . . . as a useful supplement to them," the novelist Fay Weldon has observed. "Bosses are as parents, subordinates as offspring, and colleagues as siblings." In such a context, sibling rivalry and rebellion against authority are inevitable—and lots of practice in coping with them is invaluable. The manager's job often boils down to this matter of resolving emotionally loaded conflicts, soothing hurt feelings, and inspiring people to do their best.

Jane Lubchenco, the zoologist introduced in Chapter 5, whose management tasks have included chairing her department at Oregon State University and presiding over the American Association for the Advancement of Science, the world's largest scientific society, says one thing she has learned from rearing two sons is the art of staying "analytical" about others' emotional outbursts. "I became much better at conflict resolution among my colleagues and at adjudicating differences, because I could see their behavior as grownup expressions of what my kids were doing," she says. "I could be less personally involved about it all, because I could say to myself, 'That's just a grownup temper tantrum.'" As Kathy Mayer, the Colorado physician/supervisor, puts it: "Before being a mom, I might have sympathized *more* than I should have. But now sometimes I'll be listening to someone tell me about a problem, and I'll think, and sometimes say, You are more whiney than my nine-month-old! Like, get a grip."

Discipline is also part of a manager's job, though that particular word isn't usually applied. Bosses, like mothers, must be good at establishing rules and enforcing them. In San Francisco, Hayes-White says her sustained practice at home with her children has been a powerful help for her in mastering that job as fire chief, a position to which she was appointed at age thirty-nine, in 2004. "I've been able to clearly articulate what the limits and boundaries are," she says. "In both situations, it's like: Here are the guidelines, these are the expectations, and here are the consequences if you violate the guidelines."

Even so simple a domestic tool as giving someone a "time-out" has proved effective at work, according to Lubchenco. "Although, of course, I wouldn't be telling someone my age to go to his room, I had gained this understanding that sometimes people need to blow off steam, to go away and have an opportunity to think about things, instead of pushing to resolve it all right now," she says. "The concept of a time-out not only lets your kids cool off, but you can, too." Other manager mothers cite the crossover value of tactics such as "catching" a child doing well, and being generous with legitimate praise and appreciation, all helpful in winning kids and employees to your side.

Emphasizing these behavioral similarities between parents and managers, reporter Crittenden tells of the "ah-ha moment" that inspired her book on manager-mothers. Shortly after her son was born in 1982, she was "busily devouring baby books," she writes, "and noticed an uncanny resemblance between the advice found in many books on parenting and the material in books on management that I had read as a business reporter." Her hunch was that much of the material was simply repackaged for different audiences. Years later, Crittenden pursued that hunch by attending a Harvard University seminar led by the management guru William Ury, on the topic of "Dealing with Difficult People and Difficult Situations." Ury coauthored the best-selling business negotiation guide, *Getting to Yes,* and authored its sequel, *Getting Past No.* As Crittenden describes it, more than 150 people, including senior managers of big corporations and military officers, were paying nearly $2,000 to attend, "and sure enough, the management tips . . . were largely the same lessons anyone could read by picking up a ten-dollar paperback on parenting."

Ury publicly attributed his advice to "such impeccably masculine sources as Sun Tzu, the legendary Chinese general . . . and Carl von Clausewitz, the Prussian military strategist," Crittenden says. But over lunch, she says, he cheerfully confirmed that much of what he taught actually came from Haim Ginott, the psychologist author of the 1956 parenting bible, *Between Parent and Child.* Ginott stresses the fundamental importance of helping your child feel he has been heard and understood—a tactic as useful for managers as for parents. Similarly, business management experts today frequently tout the values of restraining their macho instincts to dominate negotiations and seeking a win-win solution.

This "cutting-edge" emphasis on building team spirit and a feeling of inclusion was championed back in the 1920s by a pathbreaking woman: Mary Parker Follett, a civic activist, author, and business management consultant. Follett lectured widely on the usefulness of "participatory management," in which managers and workers would view themselves as partners, and advocated something similar to what we currently refer to as the win-win approach in negotiations. Although they were embraced in Japan and other countries, her ideas were ignored for several decades in the United States.

Back then, the idea of management as a formal field of study was just taking shape, and the dominant model was the kind of hierarchical organization inspired by the military, with men holding virtually all the executive positions. Today, however, Follett is once again in vogue. Sir Peter Parker, chairman of the London School of Economics, has suggested that Follett, who had no children of her own, be thought of as the "mother" of management.

The general idea that *female* managers often succeed with a more "feminine" and "participatory" style has been the topic of several recent books. In 1995, Sally Helgesen's *Female Advantage: Women's Ways of Leadership* profiled four women business executives who followed a "feminine" model of direct communication, valuing relationships, and a "web of inclusion" to achieve success. Another book, *Why the Best Man for the Job Is a Woman: The Unique Female Qualities of Leadership*, followed fourteen executive women who "tap into their feminine side to lead." "It's partly team building," one woman executive told Sumru Ekrut, the director of the Wellesley study of female leaders. "And a family is partly team building too. Getting kids to work together and feel the family feeling and not to be hitting each other and so forth."

Mothers indeed must constantly subjugate their own wishes as they figure out what will make most of their family the happiest. "Whether it's putting off a phone call to respond to a baby, or cutting someone slack with their homework so you can all get to the park, you just get much better at various junctures of the day in deciding what's best for the group," says San Francisco working mom Sue Kostal. "After a while it becomes pretty natural to figure out without a lot of agonizing what's the most painless way of accomplishing something."

The team-building concept extends to another basic manager's job: mentoring. Mothers who've become practiced in coaching children to do their best often find themselves less self-conscious when doing so with others. "If you watch a mom taking her kid to the park, it's really a developmental experience," says Rayona Sharpnack. "She's telling the child to stop and wait for the light, and to look both ways before crossing the street, etc., so when she gets to work, this training and developing practice is already second nature."

One managerial trait reportedly common among working mothers and demonstrably appreciated by their employees is a resistance to micromanaging. Mostly this is because they just don't have the time. But that deficit translates to strength when executives know when to delegate tasks, which helps build strength among their employees. Besieged working-mother managers might at first delegate out of necessity, see that it works, and then do so more strategically the second and third time around.

After six kids and seven years as CEO of the company she founded, Trainer Communications in Danville, California, Susan Trainer realized that stepping back would be good both for her family and her firm. "The strategy that had helped me grow the business to its current level wouldn't work going forward if I insisted on being so hands-on," she told *Working Mother Magazine.* Trainer began delegating major tasks to her staff, stopped leading every meeting with clients, even ceased attending some meetings with new customers. "It needed to be that way," she said. "We can double the company in size again, but not by my working harder."

Caregiving: On the Job

Maria Calles, an immigrant from El Salvador, works in what would initially seem to be a wholly different job universe than that occupied by Susan Trainor. As a health aide to home-bound elderly people in the San Francisco Bay Area, Calles holds the kind of service job that has exploded in growth as people in industrialized countries live longer, have fewer children, and often must depend on strangers to care for them. Tending to a dependent adult may seem like mentally simple work, but it can often demand a lot of brainpower, as Calles knows well. Through caring for her teenaged son, Fernando, she has realized the importance of reading her clients' states of mind, respecting their boundaries, and restraining her natural impulse to, as she says, "be too much of a help." From time spent with Fernando when he was in emotional pain, she says, "I have learned when to step in, and when I must leave him alone. You learn these things well with your children, because they are very direct with you when they set limits." As always, intensive practice in certain behaviors can change the way you think, which can change the way you behave in the future. Working mothers who care for their own children for large parts of each day may accordingly bring special talents to the millions of jobs available in fields such as nursing, physical therapy, and fitness.

But to be sure, talents in empathy and other social skills so relevant to caregiving jobs are certainly valuable in jobs with broader general concerns. Although it never appears on an asset inventory, the caregiving skill of a company's managers can often make a subtle but critical difference between high or low morale and productivity. "We may not notice the amount of caretaking done by an artist with apprentices or by a chief of engineering or a college president; we fail to notice the aridity of these jobs when they do not involve care for others," writes Mary Catherine Bateson in *Composing a Life.*

After my psychiatrist sister Jean switched from seeing private patients to managing a large staff of doctors for Colorado Permanente in Denver, she was surprised by how much of her job involved mentoring, resolving conflict, and otherwise mothering other employees. "It can take hours of my time every day," she says. Barry Lester, a pediatric psychologist in Rhode Island, says experience with babies—your own, or other people's—can make a big difference in how well you do this. "From general experience, having kids . . . makes you more understanding of other people, and more patient," he says. "You're able to stand back more and maybe have a broader perspective and not rush to judgment."

Lester believes that, in particular, mastering the ability to read a baby's nonverbal signals can be a great step forward in developing successful relationships with adults. He has seen this connection with his own roles as a father of three and head of his clinic for colic, and he has observed it in his fifty-five coworkers, whether or not they have children of their own. "They get very good at picking up stuff from other people because they're always doing that with the babies," he says. "They are exquisitely sensitive in ways in which they relate to each other."

"People ask me, would you be more likely to hire a mother?" says Hayes-White, the San Francisco fire chief. "I always look at overall competence, but I do believe it's an added bonus to have someone who's had the experience or is experiencing being a mother." Hayes-White describes having parents on her fire-fighting force as a "service to the community," not only because it increases diversity but because she feels parents might naturally be more emotionally responsive and essentially better at dealing with other people. "You change. Your life is no longer centered on yourself. You have to adjust to more responsibilities, and I think you get a broadened perspective."

This perspective is useful in various ways for a fire-fighter, she says. "It's a family type setting here. We care for each other. When you're working a twenty-four-hour shift, you're sharing meals and common dormitories. And then, there's a lot of caregiving to others in this job. We render medical care, and lots of times you're dealing with very emotional situations. When you have to notify families of negative consequences to a loved one, your compassion skills become very important."

As Sara Ruddick pointed out twenty-five years ago, the practice of caregiving can lead to a distinctly different way of looking at the world, including one's professional sphere. Many working mothers envision their caregiving experience at home as a well of creativity when it comes to their work. Writing for the *Columbia Journalism Review,* Judith Matloff, a former Reuters News Service foreign correspondent, cites several other foreign-correspondent mothers who felt that parenthood made them better reporters with sharper insights into human suffering. "I have taken many incredible pictures of women and children in war and in hunger that made an impact on world opinion," said CNN's Cynde Strand, supervising editor on CNN's international desk. "But I never really saw those pictures until I had a child of my own."

Motherhood can also provide a deep and immediate connection with others in more prosaic settings, adding emotional grease to customer loyalty. At Mollie Stone's grocery store in Greenbrae, California, for instance, shoppers will often choose checkout clerk Patty Ochoa's line, even if other counters have less of a wait. They're eager to commiserate about their kids, and hear the latest on Ochoa's teenagers, she says. "There is a huge number of moms who pass through here, and you have a certain kind of conversation with a customer who has kids," Ochoa says. "You get a special rapport, because the single women (clerks) just can't go there." This social advantage can cross class lines. "Let's face it, most senior executives have kids," says Kathryne Lyons, a commercial real estate broker in Manhattan. "So when I had my baby, it became a common point for us to talk about. Several of my clients have sent me flowers, and most of them ask about my daughter when they see me. It just gives us something special to talk about."

The mere realization that others have multidimensional lives can be transformational for some women, or men, who've been able to keep their focus fixed on the progress of their careers. If you have limited other outlets in your life, you may tend to grasp your work too closely, making small things into big things, and end up being a pain in the neck to your coworkers. Before I had my family, I could never quite understand why my editors kept complaining about my constant fussing with my stories as I attempted rewrites long after the official deadline had passed. I self-righteously felt they resented my perfectionism, when the poor souls just wanted to get home to their families. Today, I still can't resist a *little* fussing, but it's nothing like before. I want to get back to my own family. Like many other mothers, I still take a lot of pride in my work, but I feel I'd be failing in a major way if I didn't invest some smarts in my children. As Rayona Sharpnack told me, at the end of an otherwise easygoing interview, "If your kids end up sucking off the system, I don't want to hear about your book."

CHAPTER
11

Smarter Than Ever

Why Motherhood Today Takes a Lot of Brains

Sometimes I feel that my mother alone really knows who I am—the furtive boy, the trespasser, the secret wanker in his room.

JAMES ATLAS,
IN *MOTHERS AND SONS: IN THEIR OWN WORDS*

MARIAN SANDMAIER KNEW something was seriously wrong with her sixteen-year-old daughter, even after two doctors told her not to worry. Darrah was complaining of severe headaches and a strange "sloshing" sensation. Increasingly alarmed, Sandmaier, a freelance writer and editor in Merion, Pennsylvania, did what tens of millions of Americans do when they have questions their doctors can't or won't answer: She took to the Internet to research and diagnose the problem herself.

Sandmaier soon discovered that her daughter was having an extremely rare neurological reaction to minocycline, an antibiotic she had been taking for a persistent skin problem. Fluid was accumulating around her brain, a condition which, if untreated, might have led to chronic pain and blindness. Darrah's pediatrician hadn't even asked whether she had been taking medication. She stopped the drug and recovered within a few days. Sandmaier later wrote about her experiences in a syndicated essay that one newspaper editor headlined: "Take Two Aspirin and Go on the Internet."

The job of mothering used to be much more physical. There were diapers to wash, more meals to cook from scratch, more clothes that required ironing. Today, the work is increasingly cerebral, forcing women to think on their feet. In an era of second and third opinions, malpractice suits and mistrust, former authority figures from doctors to schoolteachers to grandmothers have lost prestige. That leaves mothers as the default option: lonely information analysts in a twenty-four-hour news cycle, our main advantage our unique commitment. We have to be smarter, and, often, we *grow* smarter, in the process.

This trend, in which human motherhood increasingly has become a more cognitively demanding occupation, has been taking place in a broader sense throughout evolution. In other mammals, parenting behavior subsides after babies are weaned, ending the strongest neurochemical influence; in humans, whose brains develop over a much longer period, it can last intensely for a decade or even two: long years in which a mother, using all her mental resources, stands between her developing child and an often dangerous world.

Chapter 10 featured how a mother's special smarts help her participate as a worker in the "face-to-face economy." This one looks at how vital they are as she navigates her children through a buyer-beware culture. Today's information onslaught requires everyone to be savvier than ever before, but mothers who must fend not only for themselves but also their children must arguably be savviest of all.

Four developments in particular have increased motherhood's cerebral challenges over the last few generations. In first place is the modern torrent of parenting theories and advice, which in recent years have had the weight of neuroscience research behind them. (Mothers are now assumed to be partly responsible for how our kids' brains are wired.) There are new and often complicated medical quandaries such as those faced by Sandmaier. (Mothers find themselves more frequently second-guessing physicians.) There's the deterioration of public schools, combined with more competitive college admissions and a much more cut-throat job market. (Mothers more than ever must be informed about and lobby for academic resources, at the same time supervising homework that can start as soon as kindergarten.) And finally, there's our culture's tremendous increase in consumerism, which threatens not only our children's well-being but the natural environment they'll inherit. (Conscientious mothers must constantly arbitrate between immediate gratification and long-term consequences.)

This growing array of responsibilities make it especially important for mothers to challenge the Mommy Brain stereotype that might lead us to think we're not qualified to outthink other authorities. Increasingly, that's our challenge. "It's continually underestimated how much mothers have to negotiate the cultures in which they live; how much they have to both live within the culture and resist it," says Sara Ruddick. This, she adds, "has always been much more complex than anyone gives them credit for and maybe today is even more complex."

"Intensive Mothering" and Brainy Babies

By the end of the 1990s (the "Decade of the Brain"), Western culture's message to mothers seemed to be that we were failing our kids if we hadn't already begun stimulating their mental equipment *in utero.* The most popular baby shower gifts included "Baby Mozart" videos, booties with bells to enhance body awareness, and "Infant-Stim" black-and-white mobiles, designed to capture and train a newborn's gaze. The so-called Better Baby movement, based at the Institutes for the Achievement of Human Potential in Philadelphia, advertised graduates such as one four-year-old who already was "reading for pleasure, performing many household responsibilities, playing the Suzuki repertoire on the violin, running in three-mile races, and swimming beautifully." Young brains, it seemed, had new and largely untapped potential; thus, whereas the 1970s message to mothers, perhaps best expressed by the novelist Philip Roth, was, "Back off, smotherer!" the new mandate was "Do more . . . and more!"

In 1997, nervous parents passed around the *Time* magazine cover story that featured neuroscientists discussing about all the little windows that are opening—and closing—in a growing baby's brain, implying deadlines for maximum benefits in exposure to foreign languages and music. The writer warned:

> In an age when mothers and fathers are increasingly pressed for time, and may already be feeling guilty about how many hours they spend away from their children, the results coming out of the labs are likely to increase concerns about leaving very young children in the care of others. For the data underscore the importance of hands-on parenting, of finding the time to cuddle a baby, talk with a toddler and provide infants with stimulating experiences.

As baby-brain cultivation increasingly became a competitive sport, a 2004 *Wall Street Journal* article hailed the advent of "Belly Talk: The Latest and Earliest in Parenting." It described a pregnant woman, "eager to get her fetus on the fast track," who was querying academic experts about how she could "maximize her baby's time in the womb," and described new developments such as a "BabyPlus Prenatal Education System," an electronic device issuing rhythmic sounds and advertised as the auditory equivalent of prenatal vitamins. Other parents tried piping in classical music via CDs and little earphones. I did this myself with poor fetal Joey in 1995, until my brother Jim suggested that Joey's ultrasound image looked as if he were screaming, "Enough of that *&^%! *Pathetique!*" He may have been right: Some experts now warn that prenatal concerts can disrupt a baby's sleep.

Still, many mothers continue to feel obliged to start parenting as early as possible, goaded as they are by new reports of the strong difference engaged caregivers can make. In the same year as *Time* published its baby-brain cover, the frenzy reached such a pitch that the sociologist Sharon Hays coined the term *intensive mothering* to describe the new standard. Hays believes the motherhood ante has been upped by the modern proliferation of child advice books, pointing out that more than 90 percent of U.S. mothers surveyed say they have read at least one of them. Although there are no comparable data from other countries, Hays says the response to her book outside the United States—especially in Germany, but also in the United Kingdom, France, Argentina, and Brazil—has convinced her that this is a nearly global phenomenon. "The more you get developed capitalism, the more you get child-rearing advice," she says. Hays believes this is partly because of the continuing march of women into the workplace at the same time as workers are becoming more mobile and extended-family support is turning into a thing of the past. "In these societies, the family becomes the last bastion of intimacy, but because there are so few relatives around to give care, the mother becomes crucial," she says. "Everyone is freaking out about how to deal with that."

The degree of our current obsession with constantly stimulating our children's brains strikes previous generations of mothers as absurd, Hays points out. It also elicits ridicule in countries such as Russia, where reliance on extended families is the norm. "After I gave an interview with Radio Free Europe, I heard that Russian women found this whole idea hysterically funny," she says. "The notion of a precise set of rules you have to follow to be a mom was really hilarious to them."

In the late 1990s, U.S. mothers were spending as much time, and possibly more, with their children as mothers had some thirty years earlier, despite the new hours most of these women were now spending working outside the home, Hays and other experts have found. For many, this has meant a big reduction in "leisure time," including sleep. And, again, it's not just the quantity of time spent with kids but the intensity of that time that has changed. "If my mom sent me out to play in the street all day, she would say, 'I'm taking care of my kid,'" Hays recalls. "But mothers today don't consider they're taking care of their children unless that time is totally child-centered." The belief that your child is being neglected if he isn't constantly occupied, engaged, and stimulated is carrying over to decisions about childcare, Hays says. Many mothers no longer trust in the simple solution of having relatives care for their children—even when relatives are available and the mothers are short on funds—but instead will seek out professionals, which requires not only more money but more brainpower to discriminate among options. Mothers with the resources to do so will also feel obliged to pile on after-school enrichment programs.

The matter of how children are disciplined can also tax a mother's mind more than it did in years gone by. Although only a generation or two ago, it was still socially acceptable, if not necessarily *right,* to spank, the more cerebrally challenging technique of negotiation is now the Western cultural standard. The switch has come about as part of a fundamentally different, post-Victorian way of perceiving children, according to Peter Stearns, author of *Anxious Parents: A History of Modern Child-Rearing in America.* Whereas kids used to be seen as sturdy and resilient, they now are perceived as vulnerable and malleable. This change has put parents squarely in the line of fire between authors of advice books who endorse permissiveness and social critics who bewail the prevalence of brats.

Once again, it's more often mothers than fathers who end up engaged in the makeshift reasoning, bargaining, and bribing that comprise contemporary discipline. Many men now resist the old standard of being marked as the bad guy, making an empty threat of the time-honored admonishment "Wait till your father gets home." And this leaves mothers, who for the most part spend more time with their children, holding the bag, or, more specifically, reading the discipline books, attending the seminars, and drawing up elaborate star charts with scheduled rewards or reprisals. "I've taken two parenting classes and tried about five behavioral programs in the last two years," says Michelle Bullard, a flight attendant and mother of two preteens. "They all work for a while, and then I have to figure out a new one. When I was growing up, no one needed to know or care what your opinion was. I didn't think that worked—we kids all had closeted resentment. But now it's the moms' therapy bills that are going up."

More Health Worries, Less Reliable Medical Advice

My parents, who thought parenting was hard enough, didn't use seatbelts, smoked cigarettes with the windows rolled up, went out on the town without cell phones, and occasionally even fed us TV dinners. I, on behalf of my children, worry about rising rates of asthma, autism, and allergies, in addition to antiobiotic-resistant disease, Anthrax, GMOs, mad cow, climate change, nuclear proliferation, and, for good measure, perchlorate, a toxic component of rocket fuel recently discovered in cow's milk.

Whereas our ancient ancestresses had but to turn their noses to the wind or listen for the sound of roars, we read newspapers, scan e-mails, analyze the latest color-coded security alert, and network like mad to protect our offspring competently. All the while, à la the Serenity Prayer, we need to summon the wisdom to tell the difference between what we can and can't change.

My peculiar temperament and journalist's access to information gives me the disposition and opportunity to be nervous about an unusually wide range of potential perils. But even the calmest mothers who never read newspapers will find they have little choice but to confront at least a few basic health threats.

A casual glance around the schoolyard may provide sufficient notice of one modern peril: the startling increase in childhood obesity. Throughout the world, children are becoming overweight in a rapidly growing trend: In the United States, the number of overweight kids doubled between the 1980s and first years of the millenium, leaving one in five U.S. children at increased risk for hypertension, diabetes, and heart problems. Adults in most developed and many developing countries have similarly been tipping the scales, a problem that the World Health Organization has attributed to a range of contemporary pitfalls, including the increasing availability of around-the-clock television, rising levels of promotion and marketing of fattening foods, and more frequent use of restaurants and fast-food outlets. Because mothers are usually their families' main grocery shoppers, lunch packers, cooks, and sports coordinators, as well as the ones who most often must yank out the computer and television plugs, our mindfulness can make a big difference here. But again, we need to stand up to a culture in which schools have cut back on physical education while installing Coke machines; elementary school teachers pass out candy to perk kids up for their standardized tests; and our children's playmates' parents drive them two blocks to school and have their birthday parties at McDonalds.

Another big change that has come just in the last few generations is an awareness of the dangers of pesticide residues in produce, a danger all but ignored until the 1962 publication of Rachel Carson's *Silent Spring*. Forty years later, millions of responsible parents were heeding warnings that such chemicals can cause birth defects, cancer, and nerve damage, and that they are particularly harmful to children. The juggernaut global growth of the organic-food industry over the past few decades has contributed to make "good mothering" expand to encompass an awareness of farming conditions. Today, there's an increasing cultural expectation that mothers pay attention to the source of our food while also reading labels to look out for nasty additives such as trans fats, the partially hydrogenated vegetable oils that increase shelf life at the cost of clogged arteries.

No longer, furthermore, can mothers guiltlessly sit back and read a magazine while visiting their children's pediatrician. Although there's debate in the United States as to whether doctors are actually spending less net time with patients under our more managed health care system, there's little question that they are generally much less trusted today than ever before. The 2004 scandal over doctors' being paid by big drug companies to write prescriptions was simply another milestone in a long decline of physicians' public stature. Mothers are now more inclined to inform themselves so that they can ask educated questions. "None of the doctors I saw were remotely as committed or motivated as I was about my child's health," recalls Sandmaier, whose daughter suffered serious side effects from her antibiotic. It was this motivation that spurred her to track down Neuroland.com, the Web site that first suggested the drug, minocycline, could be dangerous. As Sandmaier later wrote, she felt blessed that she had the means and the know-how to use the Internet when millions of other parents had no such chance.

In fact, however, mothers from a range of social classes have now become more Internet-savvy, and it is estimated that more than 40 million Americans search the Web for health information. Anna Branscum, a San Francisco Bay Area drugstore clerk, became one of them when her pediatrician dismissed Branscum's concerns about her daughter's weight.

"She said I should just watch it, and do the best I can," she told me. Instead, Branscum tracked down a site called Blubberbusters.com, which provides links directed to children and preteens, offers tips about how to eat less, recounts success stories, and offers chat-rooms for conversation with other heavies. "At first I thought, 'Right, she's going to go for this?'" Branscum recalled. "But it spoke to her; it turned out to be helpful."

Another pervasive health dilemma sending mothers in droves to the Internet is that of whether to follow a teacher or doctor's recommendation to medicate a child who seems depressed or spacey at school. The practice of medicating children, which began in earnest in the mid-1990s, has become so widespread—attention-enhancing Ritalin production soared by 500 percent between 1990 and 1995—that it's easy to believe you might be letting your child down by rejecting pharmaceutical support. Doctors and teachers often point out that a child's self-esteem might suffer if he or she can't keep up with schoolmates. On the other hand, researchers have uncovered serious side effects of antidepressants and stimulants given to children. In the end, parents have little choice but to seek a physician they trust, rely on their own intuition and, as always, try to be informed and courageous enough to stand up to cultural trends.

Take-Home Lessons

I was sitting at my kitchen table the other day with my friend Idie Weinsoff, an artistically gifted fifth-grade teacher, while she counseled me on the potential design of a three-dimensional model of a black hole in space. My son Joey, who chose the topic for his third-grade report, was happily trading Pokemon cards in his room with a friend. "What are our children learning from these art projects?" Idie asked. She then supplied her own answer: "They're learning that their mothers will do them." In 2000, close to 60 percent of American parents were helping their children "considerably" with their homework, some explaining that it simply wouldn't be done otherwise, according to *Anxious Parents* author Stearns. The sheer quantity of homework has moved some cities, such as Alexandria, Virginia, to pass laws limiting the amount that teachers can assign. At many elementary schools, parents also complain that the intellectual level of after-school assignments has risen beyond not only their children's reach but their own.

Mothers are frequently left with the no-win choice: Do we take on our kids' homework as our own, and sap their industriousness, or do we force them to tackle research reports, complete with footnotes, at age seven, and sap their self-esteem? The "fairy-tale" puppets I saw at the last second grade open house gave me a clue about what most of my mommy peers were doing. Although I'd potentially humiliated my own son for life by turning in a sock dragon with little paper wings and button eyes sewn on, I suspect the other moms had sought the aid of professional crafts consultants to produce their elaborate papier-mâché models. Or else, like Idie, they were all geniuses.

No small part of the after-school load, which most mothers must confront after their own hard day's work, is helping children prepare for exams, including statewide standardized tests that start early in grade school. Tough standards from politically pressured public schools in turn raise the pressure on mothers to second-guess the school system. Some who can afford an extra year of preschool hold their children back as late as possible before entering kindergarten. Some mothers won't permit their kids to participate in rigorous statewide tests. And some determinedly investigate the range of special resources for potential handicaps that currently include speech problems, learning issues, and even, at my children's school, difficulties with "sensory awareness in space." In the San Francisco County School District in 2004, more than 20 percent of the approximately 56,000 children enrolled had been designated as requiring special services for any one or more of a wide variety of physical, learning, or emotional disabilities. The services provided included resource specialists, separate classes, occupational therapy, speech and language support, and mobility assistance. If your child has any such issues, and his teacher hasn't already spotted them, it's your job to figure them out and seek the extra attention, at the risk of letting him down.

But let's assume your child survives the school system and even does well. Now your main brain teaser is to help prepare him for college and a job in a competitive new world economy where downsizing has considerably up-sized the job of a mother. Increasingly, parents have to summon their creativity to make sure their kids not only accomplish high grade point averages but distinguish themselves by roughly the age of eight by their precociously altruistic and creative extracurricular interests. "They're always raising the bar," says Carole Gifford, who worries that her preteen son still has few interests other than video games. "You used to be able to get into a University of California college fairly easily, but now you've got to have the 3.9 GPA, 1,300 SATs, and have built a church in Nicaragua."

As if the stakes weren't high enough, contemporary industrialized-country parents know that competition for jobs will only increase in an increasingly globalized economy. My mother used to tell us to eat all our vegetables because starving children in China and India didn't have that luxury. I tell my children to study harder, because otherwise children in China and India may grow up to buy and sell them.

Buy, Buy, Baby

Compounding the modern cerebral challenge of raising a healthy and well-educated child are the dogged efforts of powerful corporations that, but for mothers, mostly, would have virtually unlimited access to our kids' developing brains. Mothers have the thankless mission, should they decide to accept it, of analyzing content, switching channels, perhaps even turning off the television, and saying no to constant demands for costly plastic and electronic devices.

By 2000, the typical American child aged from two to seventeen was spending on average just short of twenty hours a week watching television, according to Nielsen Media Research. This is in contrast to the reported 38.5 minutes a week that average parents spend with their children in meaningful conversation. And even if you're in the tiny minority of families that don't have a television, you may find that your preschool, or even your elementary school, doesn't share your values. One 1994 study found that 70 percent of daycare centers use television during a typical day.

Television can hurt your child in two main ways in addition to preventing him from doing something healthier, such as exercising or reading. The first is by exposing him to an extraordinary amount of violence. A 2003 Kaiser Family Foundation study reported that almost two out of three television programs contained some violence, averaging about six violent acts per hour. Marco Iacoboni, the UCLA brain-scanner, is one of several brain researchers who think we should worry a lot more than we now do about such statistics. He says brain research (including new findings about "mirror neurons," described in Chapter 8) strongly suggests that a child who watches repeated violence will be more likely to commit violence himself. "Seeing violence over and over again facilitates the mimicking of such acts, because you activate the same motor regions as if you were committing that act," Iacoboni says.

The second main way that television harms kids is by advertising that encourages what are historically unprecedented levels of consumption. Much research suggests this trend is unhealthy both for children's values and for an environment overwhelmingly taxed by resource extraction and waste disposal. By 2004, spending for advertising and marketing aimed specifically at children had risen to more than $15 billion a year, and the average child was watching more than 40,000 television commercials annually. One poll that same year found that 87 percent of Americans thought the so-called consumer culture made it harder to instill positive values in children. But in an unrestrained market economy (compare ours with that of Sweden, where it's illegal to direct ads at children younger than twelve), it remains the job of parents not only to limit exposure to television and analyze its messages but also to encourage other pursuits and put brakes on kids' desires for more "stuff."

Face Time

At a time when the great majority of U.S. mothers are working outside the home, one of the most taxing mental challenges you may face is to figure out how much and when your children truly need your company and how you can manage to be there. Answering these questions effectively in the context of your families' own circumstances requires you to be informed about the current state of knowledge pertaining to children's needs, and, once again, often obliges you to stand up to cultural pressures and hype. The main difficulty is that there has been such sharply conflicting information in recent years. To paraphrase a bumper sticker: If you're not confused, you haven't been paying attention.

If ever the issue of what babies need was intellectually demanding, it became much more so with the popularization, beginning in the late 1960s, of the theories of the British psychoanalyst John Bowlby. In his landmark trilogy of books and public speeches, Bowlby stressed what he described as a genetically programmed need of human babies, similar to that of other primates, to "attach" to a single trusted figure, preferably the mother. The quality of this attachment, he warned, would be key in determining a child's subsequent emotional development. Bowlby's theories became well known at roughly the same time that millions of U.S. women were contemplating whether to take advantage of new opportunities to join the workforce. Women who had genuine freedom to choose—that is, whose paycheck wasn't essential to the families' finances—felt obliged to weigh the impact of their going out to work on their children's well-being. Women who could not afford to choose suffered particularly painful guilt.

Yet, some thirty years later, mothers could find relief in a hugely controversial book titled *The Nurture Assumption,* by Judith Rich Harris. A grandmother and former textbook author, Harris cited numerous studies showing that parenting is by far a lesser long-term influence than inborn temperament, birth order, and peers. Research on identical twins raised apart, for example, suggests that about 50 percent of adult personality traits are decided by genetic factors and that being raised in the same home seemed to have minimal effect on twins' personalities. In her interviews, Harris was even more emphatic than in her writing. Many parents are reluctant to have kids, she said at one point, because they mistakenly feel that motherhood requires such a huge commitment. "If they knew that it was O.K. to have a child and let it be reared by a nanny or put it in a daycare center, or even to send it to a boarding school, maybe they'd believe that it would be O.K.," she said.

The *New Yorker* magazine called Harris's book an "utterly persuasive assault on virtually every tenet of child development." Yet the relief she hoped to offer didn't last long. If anything, there has been an increasingly fierce debate in the early years of the new millennium over the minimum mothering that children need, and scientists are entering the fray more determinedly than ever before.

In September 2003, the Commission on Children at Risk, a panel of academically distinguished scientists and physicians joining with YMCA officials, released a report that highlighted escalating cases of emotional and behavioral problems among children, including depression, suicidal thoughts, violence, and anxiety. (It alleged, for instance, that 21 percent of children aged from nine to seventeen "had a diagnosable mental disorder or addiction.") The panel, which called its report "Hard-Wired to Connect," went on to detail the scientific basis for children's need for strong attachments to others, a need the authors maintained is going largely unmet. Although the authors didn't explicitly call this quandary the responsibility of mothers, the report highlighted animal research, including that of Stephen Suomi, a psychologist with the National Institute of Child Health and Human Development, who has demonstrated that strong maternal care can improve the brain functioning of rhesus monkeys genetically predisposed towards high levels of anxiety, aggression, depression, and substance abuse.

The television program "Good Morning, America" called the report "a real wake-up call for America's parents." And the Motherhood Project, sponsored by the conservative Institute for American Values, issued a statement saying that the commission's findings "should give mothers the strength to say that mothers are central to children's development; we are not easily interchangeable with other people. These findings should give us the courage to declare that bonding is a feminist issue and a human rights issue."

Suomi has not been alone in turning up persuasive evidence about the life-long value of good mothering—although it remains scientifically unclear that the mother *herself* has to provide it. In a fascinating report in 1997, researchers studying rats showed that pups who were enthusiastically licked and groomed by their mothers grew up to be much more resilient and less anxious adults.

Research on humans has also turned up compelling evidence of the influence of parenting on children's response to stress. Since the late 1980s, Mark Flinn, a professor of anthropology at the University of Missouri, has been studying almost three hundred children living in the rural Caribbean island village of Bwa Mawego. In one of the most extensive, long-term human studies to date, Flinn has taken more than 25,000 saliva samples from these children, checking for the stress hormone cortisol, while also monitoring other measures of health and daily events in the children's lives.

From his findings, Flinn argues that the quality of home life is a fundamental determinant of children's mental and physical health; and, contrary to Harris, he's convinced it's a much more important influence than that of peers. A child, he says, can walk away from a significant conflict with playmates with little change in cortisol levels, yet when that same child is scolded by a parent, his stress hormones shoot up. "There's nothing more important to a child than figuring out what makes those close to him happy or sad," he says. Flinn has also found that kids who live with both biological parents have lower average cortisol levels, weigh more, and grow more steadily than those living with stepparents or single parents unsupported by relatives.

Few mothers can read this research without reflecting on their own parenting. Are we licking and grooming enough? Yet for most of us, it's also apparent that the world in which we parent is significantly more complicated than cages in a lab or even the village of Bwa Mawego, obliging us to make nuanced choices between our children's emotional and material needs, and our own. Ravenna Helson, the Berkeley sociologist who has studied Mills College graduates since the 1950s, notes that more than half of them said they were depressed when their children were young, a condition she thinks was the result of their not having jobs, or even a notion of how their future would unfold after their children grew up in an era when other women were just beginning to explore historically new possibilities. Today, says Helson, women facing the conflict between jobs and children feel less depressed, although they may be, as she put it, more "cross."

It seems unlikely that, conflicted as they may be, mothers will march back in droves to their homes, surrender their financial security, and trade "cross" for "depressed." Instead, we'll likely keep straining our brains to find ways to fit everything in. On Mother's Day in 2004, the *New York Times* published a chart that contrasted the lives of mothers who work outside the home with those who don't. The employed moms had substantially less total free time, watched less television, slept less, and, in comparison, said they "always feel rushed," yet when it came to their overall well-being, 85 percent of them, versus 77 percent of the stay-at-homes, said they "get a great deal of satisfaction" from their family lives.

If, as many studies suggest, depression in mothers can lead to depression in children, it may be reasonable to expect that more satisfied mothers will have happier offspring, especially given that we seem to be spending just as much time with our kids as our mothers did. At the same time, many mothers worry with good reason that their jobs are keeping them away from their children at times when they need us most. As I discuss in the next chapter, finding a work schedule that allows you to take parenting seriously can become one of the greatest mental challenges you face. But some smart employees and bosses have been making progress in the quest for solutions.

Part Four
NOW WHAT?

CHAPTER

12

Reengineering the Mommy Track

Ideas from Some Brainy Travelers

Mothers' passionate concern with keeping beloved children alive is a central fact of life—so primary that it is taken for granted and mostly goes unnamed—just as when we mention daylight, we assume the sun.

JANNA MALAMUD SMITH

JANE LUBCHENCO IS one of the most successful female scientists of our time, and her career as a zoology professor hasn't forced her to be any less successful as a mother. Early on, Lubchenco and her husband, Bruce Menge, also a college professor, passed up positions at Harvard, opting instead for a rare chance to split a single full-time teaching job at Oregon State University. The arrangement allowed them to share the parenting of their two sons, while freeing Lubchenco from worries that might have diminished her considerable smarts on the job.

Harvard's loss was Oregon State's gain. Lubchenco went on to shower her new employer with prestige. Her research in tidal pools along the rocky Pacific Coast won important grants and awards; her graduate students achieved faculty positions of their own, and she served as the first female president of the American Association for the Advancement of Science. Meanwhile, Lubchenco breastfed each of her children for more than a year, coached their baseball and soccer teams, and, together with Menge, gave students a striking model of how to excel in science without short-shrifting families or relegating women to a career-derailing "mommy track." "The great benefit I didn't even consider at first was how *happy* they were," says Frederick Horne, the former dean of science who supervised the couple and several other job-sharing couples. "There's no question that it made them better workers."

For a mother to muster all her smarts at work, a certain amount of "happiness" at home is fundamental. The minimum standards for most women mean knowing that her children are safe and well cared-for, having enough time with them to feel like a competent parent, and having it all, somehow, be economically affordable. "There's this treasure just sitting out there for employers in all the moms with so much to give who would work so hard and well if they knew they could also do right by their children," says Catherine Gray, the working-mother president of The Natural Step environmental group.

This condition seemed rather too tall an order in the first few years of the new millennium, when a job slump created a buyer's market for talented workers. Nearly one-third of U.S. companies in those years were downsizing their family-friendly programs, such as job-sharing and telecommuting, arrangements with which they'd previously wooed employees. Workers lucky enough to have jobs were putting in more hours: One report shows that dual-earner couples with children under eighteen toiled an average of ninety-one hours a week in 2002, about ten more hours a week than in 1977. And they were doing so even amid reports of a significant decline in children's mental and physical health. Critics blamed working parents for what seemed a certain toll on society at large. "There's trouble in the engine and we're junkyard bound if some moms and some dads don't start hanging around," wailed the folk singer Iris Dement.

Some mothers apparently took heed of these warnings, or simply decided that clinging to a job while rearing kids was too hard. As noted in Chapter 7, the U.S. Census in 2002 recorded the first decline in labor force participation of mothers with infant children since 1976, a drop from 59 to 55 percent. Yet the heavy media coverage of this development risked obscuring larger trends unlikely to be soon reversed, and constituting continued pressure on companies to accommodate working families. In 2002, a full 72 percent of mothers with children aged one and older were still active members of the labor force. And by the year 2000, for the first time since the U.S. Census Bureau began tracking such numbers, families in which both members of married couples with children were working became the majority. They joined more than 10 million working single parents.

In those same years, millions of future working mothers were moving through the academic pipeline, earning degrees to help them qualify for good jobs. Women made up half the student bodies in most law and medical programs, and at least half the doctoral programs in social sciences and humanities, compared to about 10 percent of such degrees in the 1960s. In bachelor's and master's degree programs, they were already outnumbering men.

Company executives engaged in heavy competition for qualified workers understood that retaining and improving family-friendly policies could help them stand out in a crowd, especially since, in another fairly recent development, entities such as the Families and Work Institute and *Working Mother* magazine were prominently keeping tabs on their behavior. *Working Mother*, for instance, has since 1985 compiled and effectively publicized its annual list of the Top 100 Companies for Working Mothers, showering praise on the most innovative firms. In 2003, Jill Kirschenbaum, the magazine's editor, contended that although some companies were cutting back on family-friendly benefits, the innovative firms on the magazine's list were expanding them. In 1999, for example, less than a third of the listed companies offered a wide array of childcare and flexibility programs, but nearly all had adopted them four years later.

For-profit firms don't make such investments for public relations points alone. Several studies published in the early years of the new millennium were showing that family-friendly programs can save companies substantial amounts by encouraging employee loyalty and productivity while reducing turnover and absenteeism. Such trends have continued to encourage truly smart managers to design more livable jobs. Lubchenco's example at Oregon State is still a rare beacon of how employees and employers can work together to minimize a mother's reasonable anxieties and maximize her smarts. But there are others increasingly aware of and striving to meet those minimum standards of security, flexibility, and affordability.

Freedom from Fear

In the Mozart Room at the Children's Creative Learning Center, Lorraine Felix cuddles dozing, seven-month-old Rachel while she masterfully maintains eye contact with six-month-old Elsa, rocking in a little chair inches away. Soft classical piano music soothes other infants crawling on the spotlessly clean pastel rug. Outside, a couple dozen preschoolers goof around on a state-of-the-art play structure. Mothers working upstairs at Electronic Arts, a multi-billion-dollar video game manufacturer, drop in at will to breastfeed their babies or have lunch with their toddlers. Like regrettably few other examples in the working world, parents using this childcare facility enjoy the peace of mind that comes from knowing that, in any emergency, "they can come right downstairs instead of getting in their cars and driving fifteen minutes," says Carol Miller, director of the center, located in Redwood City, in the heart of California's Silicon Valley. They also have the luxury of knowing their children's caregivers are unusually well-qualified and well-compensated. In short, this is childcare heaven. And naturally there's a long waiting list.

If mothers have yet to convince employers of their unusual smarts, it could be that a basic fear has gotten in the way. Worrying about your baby's welfare, when your childcare situation is less than ideal, can be a debilitating distraction on the job. And sadly, for most working women, their childcare options are *much* less than ideal. Electronic Arts, which subsidizes its two-year-old childcare facility with free rent, aggressively uses it as a recruiting tool. "Human Resources will knock on my door at the last second and say, 'Omigosh, I want to hire someone! Give them a tour!'" says Miller, standing barefoot with one-year-old Jonathan snuggled in her arms. A mother herself, she points out that even though EA is 70 percent male, the workers using the childcare service are 70 percent female, a sign that mothers are more attuned to the need for high-quality childcare. "Occasionally, moms whose husbands work at EA call me up and say, 'I only just heard this was available!'" she says.

Anxiety about childcare goes with the job for most working mothers, though the degree certainly varies. When vetting preschools for her daughter, Jennifer Griffin, a Jerusalem-based reporter for Fox Broadcasting Company, had to relegate concerns about teacher/student ratios and discipline policy to second place; whether there was adequate protection from suicide bombers came first. After ruling out a site where a bomber's head had recently rolled into the schoolyard, she lobbied, with other parents, to hire a front-entrance security guard. Still, even women with considerably more prosaic jobs can justifiably work themselves up about the general state of childcare, particularly in the United States, where the median hourly wage of caregivers compares to that of burger-flippers, where turnover averages about 30 percent a year, and where Electronic Art's (EA) exemplary staff/child ratios of above 1 to 8 might be achieved at best during flu season in a snowstorm.

This preoccupation has become a kind of elephant in the boardroom. With too few realistic solutions that might allay a mother's fears, she and her employers often simply try to avoid the issue and hope that she can work as if she were an unattached man. The mother suffers, her family suffers, and sometimes her work suffers as well. As the psychiatrist Paul MacLean has observed, what makes being a mammal so painful is being separated from its own kind. Mothers know this from their children's howls during time-outs, and from the catch in their own throats as they pry their toddlers' fingers from their legs on the first day of preschool. In a wide variety of animal species, mothers and offspring show distress at separation, with the effects as perceptible as the higher levels of stress hormones circulating in their blood.

In humans, however, this distress can be greatly reduced by the knowledge that someone you trust is watching your child. "It gives you huge peace of mind to know they're in a place that you feel good about," says Lubchenco. "You don't have those nagging worries, and that allows you to focus." This same, all-too-rare peace of mind has also been key to Joanne Hayes-White's extraordinary career as a San Francisco fire chief. Not only is her retired husband a willing "Mr. Mom," but Hayes-White lives next door to her parents. "I work so much better knowing that my children are well-cared for," she says. Still, it's not always easy. Hayes-White remembers when, after a five-month maternity leave, she returned to work and had to stay there for a twenty-four-hour shift. "I must have called a dozen times," she says. "Like, what am I missing? What is he doing? Then I got back and looked at him and said, 'It just looks like he's gotten so much bigger!'"

Accommodations such as EA's onsite childcare remain unfortunately rare, even though they are treasured by workers and managers alike. "It can and does give people the ability to focus on what they're doing at work," says Curt Wilhelm, the company's director of facilities and corporate services. In one U.S. survey, a full 49 percent of employers reported that company-provided childcare services help boost employee productivity, according to a report by the Child Care Partnership Project, sponsored by the U.S. Department of Health and Human Services. Offering care for children when they're down with the flu can also save managers money, another study has shown. An average business with 250 employees can save up to $75,000 per year in lost work time, researchers found.

Similar bottom-line assessments have helped inspire IBM's impressive family-friendly services. The giant computer firm has won a place on *Working Mother*'s top ten list for nearly twenty years, and been judged best in class in providing childcare. The company not only offers scores of full-time, on- or near-site childcare centers, serving more than 2,300 children, but sponsors care for school-aged children before and after classes and during school vacations.

Forward-thinking companies investing in working parents' sense of security can also gain in terms of the more intangible benefit of bolstering employees' self-esteem. As Ravenna Helson, the Berkeley sociologist, has noted, a feeling of competence at mothering can give a woman more self-confidence in confronting other tasks. Furthermore, a mother who trusts that her employer is helping her protect her child will be exceptionally grateful, a gratitude that often turns to loyalty. "I love this company's attitude," says Emily Kenner, a senior product manager at Electronic Arts whose infant son was cared for at the Creative Learning Center. Kenner, who recently transferred overseas, had a choice of going to London or to Prague. She chose Prague specifically because EA has an on-site daycare center there.

Flexibility: The Gift of Time

Once you know that your children are safe while you're at work, your next priority will probably have to do with the quantity and quality of time you spend with them. (Rarely does public discourse pay heed to the "aching desire to be with their children that many mothers feel," maintains Daphne de Marneffe, a clinical psychologist and the author of *Maternal Desire: On Children, Love and the Inner Life*.)

Can you be there when they need you the most? Can you make it to your son's poetry reading, take your daughter to the pediatrician, even tuck them in at night? The ability to time all these occasions with skill is another great contributor to a mom's feeling that she's doing a competent job as a parent, and it all depends on what's known as flexibility. Flexibility doesn't come without strings. It's usually not free time, just permission to accomplish the same amount of work outside of normal working hours. In an era of soaring employee complaints about work and family conflicts, much research has shown that when managers *can* grant flexibility, it's smart to do so. Employees tend to suffer less stress, be more productive, work better independently, and stay focused on results. Once again, they'll also have cause to be more loyal, which can translate into substantial savings.

In particular, smart bosses have found that granting working parents flexibility can reduce costly absenteeism, because few jobs—in a mother's mind, at least—are more inflexible than caring for a sick child. Family issues, which can include children's health, account for about 24 percent of unscheduled absences, according to CCH, Inc., a leading provider of human resources information. In 2002, CCH estimated the average annual cost of absenteeism at a record high of $789 per employee. These costs accrue from overtime for other employees, lost productivity, and low morale.

There are data to show that flexibility can also improve workers' output. In one survey, 56 percent of employers rated their "flexible" staff as more productive than other workers. There can be many reasons for this. Working from home can cut down on distractions and spur employees to focus their efforts so that they can finish work earlier. Compressed work weeks, offered at IBM and other big firms, have the same rationale.

Despite evidence of the advantages of flexible working arrangements for bosses and employees, many working moms hesitate to ask for special consideration because of the lingering "mommy track" stigma. And some are right to do so. One study of 11,815 managers by researchers at the City University of New York found leaves of absence were associated with fewer promotions and smaller wage increases. "The old rules too often apply," said a work/family industry expert who asked for anonymity because, as she told me, she preferred to appear more upbeat for the women she counsels. Yet she added: "If you want to advance, you don't do flextime."

With a few rare exceptions, among them Jane Lubchenco, the toll on a woman's career when she opts for a slower, saner journey is particularly obvious in academia. At UC Berkeley, despite the equal shares of women and men earning doctoral degrees, women in the early years of the new millenium represented less than 25 percent of tenure track faculty. The rest occupied a "second tier," including part-time researchers and lecturers, with no job security. Moreover, these proportions had been stagnant since the mid–1990s, according Mary Ann Mason, who in 2004 was writing a book tentatively titled *Mothers in the Fast Track: The Unfinished Revolution.* "I am trying to make this optimistic, but there are a lot of discouraging data," says Mason, who fears that academia is losing "many of the best and brightest of our Ph.Ds" because of family concerns.

A big part of the problem, as Mason's research points out, was likely due less to biased department chairs than to working mothers' domestic burdens. An unusual trove of longitudinal time-use data collected at UCB showed that faculty women with children were working an average of ninety-four hours per week, fifty-three of those hours on their careers, the rest on domestic tasks. In contrast, men with children were working eighty-two hours a week in total, fifty-six of those hours being spent on their careers. Achieving greater parity at home is a battle working mothers must engage in with their partners, and often too with their own fierce desire to be first in line, all the time, for their children. Big businesses, including academic institutions, can't force that change. But, in the search for more diversity on campuses and in corporate offices, some managers have been trying to address the time-crunch problem with structural changes that might help prevent mommy-tracking from leading to career oblivion.

Berkeley, for instance, has been experimenting with a year-long "stop the clock" procedure for tenure-seeking women with a new baby, and with a modified duty option, in which new parents can temporarily scale down their teaching requirements without prejudicing their futures. And although Harvard University may have erred, decades ago, in letting Lubchenco get away, it has since adopted so many family-friendly programs that it now figures among *Working Mother*'s best companies. Harvard mothers-to-be who have worked there for at least one year now have eight weeks of maternity leave at 70 percent regular pay. The university also has six onsite childcare centers. And in a particularly innovative program designed to help pressured junior faculty combine work and family, the medical school offers fifty fellowships a year that give scholars a real gift of time: "mini-sabbaticals" which they can use to write a grant, finish critical research, develop new curriculum, or prepare a manuscript.

Outside academia, other large employers are also wrestling with the issue of how to grant flexibility without prejudicing careers, and some are having notable success. The accounting firm Deloitte more than doubled the number of employees on flexible work schedules in the 1990s, and more than quintupled the number of female partners and directors (from 97 to 567). Other companies have found they can alleviate female *and* male employees' frustration with excessive workloads without taking a financial hit. Marriott International Inc., which has held a place on the *Working Mother* list for fourteen years, offers its 121,000-member staff flexibility options of such generosity that health benefits kick in with as little as ten hours work a week. In 2000, the chain launched a pilot project in three hotels in which managers worked five hours less each week without sacrificing profits or customer service. The percent of managers who had said their jobs were so demanding that they couldn't take care of personal and family responsibilities dropped from 43 to 15 percent.

Marriott has also made gains in diversity by means of an unusual commitment to help women rise through its ranks. In 2004, nearly one third of the chain's managers were female, a standout rate achieved partly by making managers' bonuses dependent on women's advancement. Yet what may be the smartest course of all for companies and working mothers interested in more humane working situations is the crafting and promotion of flexible strategies that both male and female employees can use. Nearly half of the employees taking parental leave at Ernst & Young in 2003 were men.

Although many smart companies are leading the way in helping employees balance jobs and families to encourage peace of mind, working mothers themselves need to be realistic about what they can and want to take on, and then they need to be assertive about asking for it, says Rayona Sharpnack, the women's leadership expert introduced in Chapter 7. "This can be hard, as it still isn't really encouraged in our culture," she says. "But as long as you think you're the puppet on the string, people are happy to be the string-puller." She remembers when her own daughter was young and Sharpnack was in charge of development for a firm in New Jersey: "I had a mantra: I don't travel. I set a boundary around myself. And within nine months, seventy-five percent of this company's new business was within half an hour's drive of my home."

When companies just aren't flexible about flexibility, a mom's best option can sometimes be to find a way to strike out on her own. Jennifer Kilfoil Lee, the mother of a five-year-old son, was working part-time with Schwab in San Francisco when her bosses asked her to extend her hours. "I knew if I didn't, I risked getting laid off, but I knew if I did, I couldn't be sane," she says. So Lee decided to quit and design her own future. She became a freelance accountant, and says she tells her clients the way it has to be: "I'm a mom, so you're going to have to let me be flexible, but I'll get it done."

Smarter Politics

Besieged parents seeking for escapist entertainment in 2004 needed to look no further than a striking piece of legislation called the Balancing Act, introduced by Representative Lynn Woolsey (D-California). The 131-page measure provided a glorious vision that even its author conceded wouldn't become reality soon, if ever. Among other things, it would provide subsidies for businesses that sponsor childcare for employees' families, and expand the number while improving the quality of existing childcare facilities. It would devote more public funding for childcare for children younger than three and children with disabilities. It would even provide paid leave for parents *and grandparents* to attend kids' sports events and plays.

The proposal received little press in a news year filled with stories about the U.S. invasion of Iraq, debates about tax cuts, and an upcoming presidential election. But Woolsey defended it in an interview with Joan Ryan, a columnist for the *San Francisco Chronicle*, as a first step so that "someday people will pay attention. . . . Look, when children and families become a priority, then it will happen." As it stands now, she said, "We make parents choose between work and family, and they have to choose work to survive."

Woolsey's expertise comes from hard experience. When she was twenty-nine years old, her husband left her "virtually penniless" with three young children. Although she had a job, she also needed government assistance for food, health insurance, and childcare. During one twelve-month period, she says, she had thirteen different childcare arrangements. "This was in the mid–1960s, when being a single mother was both an economic burden and a social stigma," she said in a May 2004 address, continuing:

> We've made some progress since then, but it's amazing to me that here we are thirty-five years later. . . the economy and the family have changed radically . . . working mothers are the rule rather than the exception . . . but public policy still hasn't caught up. We have little meaningful child care assistance; we have no paid leave; in fact, we have no respect for the challenges of raising a family today; we have no recognition that there is no more important job than parenting.

Indeed, the United States falls way behind the rest of the developed world when it comes to giving struggling parents more than lip service. Only two rich nations, the United States and Australia, still lack government-guaranteed paid parental leave. Our mostly substandard and oversubscribed childcare facilities resemble Woody Allen's famous joke about the food at a Catskill Mountain resort: "Really terrible . . . and such small portions." Moreover, we have no national program to support mothers who breastfeed, despite the proven contributions to society of this practice and its positive investment in the health of future generations. Progress in any one of these areas could go a long way toward helping mothers be less anxious on the job. Such progress could also help pave the way for smarter, healthier employees years from now, as many child advocacy groups have argued. But to date progress has been slow and sparse.

Thanks to the Family and Medical Leave Act of 1993, signed by President Bill Clinton, U.S. employees blessed with a newborn today have a right to twelve weeks of unpaid, job-protected leave during any twelve-month period (Australia grants up to fifty-two weeks). Yet the law leaves many moms behind. Not only must you be employed by a company with fifty or more workers to be eligible, but even most workers who *are* eligible simply can't afford to take time off without pay. While society carries on a passé debate about whether mothers "should" or "shouldn't" be working, these mothers are at the mercy of the marketplace, meaning millions of the newest Americans have at very best a few weeks to bond with their mothers, compromising this fundamental relationship.

In 2003, President George W. Bush revoked experimental regulation that had allowed states to help these parents out by using unemployment funds to cover their wages while on leave. But more than two dozen states have been studying the viability of using their own funds to pay for family leaves. California became the first state to pass such a law in 2002, providing up to six weeks of partial-wage benefits for workers to bond with a new child or to care for a seriously ill relative.

With government-paid family leaves for births and emergencies still a futuristic dream outside the Golden State, a few savvy businesses have nonetheless gone ahead to confront another pitfall for working parents. This one involves their children's education. Although research has shown that family participation helps students succeed academically, with fewer absences, better grades, and more college attendance, many working couples can't find the time to be involved. One survey found that 25 percent of employees have trouble even getting to their parent-teacher conferences. That's why it's so impressive that one family-owned hosiery company in North Carolina arranged to bring school counselors to their facility four times a year in company-provided space, to hold meetings with parents. Another small company in Vermont gives employees time off to accompany their children on the first day of school. Politicians might take note of these few points of light.

Still, a much more urgent political priority should be the broader one of upgrading and expanding the U.S. childcare system. Today, the mothers of three in five preschoolers are in the labor force, according to the Children's Defense Fund, leaving millions of working parents struggling with the absence of high-quality, affordable childcare, pre-kindergarten programs, and after-school activities. Low-income families' situations are the most precarious. Most put their children in the care of relatives or cheap, unregulated daycare homes. Two in five disadvantaged preschoolers eligible for the highly praised but short-funded government-assisted preschool program Head Start don't participate. In many counties, state and privately funded groups offer scholarships to help struggling families find childcare, but many are overbooked. You'd have to wait two years, for example, to find such help in Prince William County, Virginia.

There used to be a strong lobby for universal affordable childcare, with many congressional bills that, unlike Woolsey's lonely gesture, seemed to have a chance of passing. But President Richard Nixon dealt this movement a debilitating blow in 1972 when he vetoed the crusading, bipartisan-supported Comprehensive Child Development Act. He said he was against committing "the vast moral authority of the national government to the side of communal approaches to child rearing over . . . the family-centered approach." Gerald Ford vetoed another, weaker attempt in 1976, and under Ronald Reagan, what was left of federal childcare funding declined by 18 percent, according to a scathing narrative by Susan J. Douglas and Meredith W. Michaels, authors of *The Mommy Myth: The Idealization of Motherhood and How It Has Undermined Women.* George W. Bush, as they noted, even proposed cuts for his own "No Child Left Behind" program.

Given this climate, it may be easy to understand why politicians have steered clear of another effort that might help working moms be at their smartest—and rear generations of smart future workers. Reams of studies demonstrate the power of breastfeeding to make children healthier and even raise their IQs. Other research suggests that employers who support breastfeeding on the job can save money from reduced absenteeism, because breastfed babies get sick less often.

Beginning in 1997, the American Academy of Pediatrics has recommended that babies whenever possible be fed exclusively breast milk for the first six months, with breastfeeding continued as a supplement to formula for the rest of the year. But most U.S. mothers never approach this standard. Instead, we have one of the world's lowest rates of breastfeeding, in part because it's so hard to combine nursing with work.

As a group, full-time working women breastfeed their children much less than part-time or stay-at-home moms. Just 12.5 percent of full-time working moms are still nursing their babies after six weeks. This is partly a matter of logistics: Some mothers simply hate it, some lack the time to settle into a comfortable routine, others rebel at having to use a breast pump, admittedly one of the most brutal mechanisms known to woman. But sometimes mothers don't breastfeed because employers or coworkers harass them. Laura Sullivan lost her job at a Michigan customs clearinghouse after her boss refused to let her pump milk on the premises. And one woman called U.S. Representative Carolyn Maloney, a pro-breastfeeding activist, to complain that although she was allowed to pump in the office bathroom, "male coworkers would actually stand outside her office and moo like cows," Maloney says.

Since 1998, Maloney has championed legislation that would signal the public health stake in having more women breastfeed their babies. It would prohibit discrimination of breastfeeding women under federal law, provide tax credits to employers who help those women with breaks from work and private rooms, persuade the FDA to regulate breast pumps and allow purchases of the often costly equipment to be tax-deductible. The commendable effort puts Maloney in Woolsey's lonely company. Yet federal government agencies and some large companies, including Amoco, Aetna, and Kodak, that by the late 1990s were offering nursing moms onsite lactation rooms and equipment have found that the effort can help keep a mother's mind on her job. As Cindy Gerbassi, a U.S. Agriculture Department lawyer told the *Christian Science Monitor:* "If at eleven weeks I had to come back knowing (my son) Michael would not get breast milk anymore it would have been gut-wrenching. I would have felt I was sacrificing my baby's health and future."

A Little Help from Friends

William James, the American psychologist and philosopher, famously observed that the deepest principle in human nature is the craving to be appreciated. Mothers are no exception. We're at our best when people around us show some understanding of the hard job we're doing and the benefits for society when we do that job well. Yet low-income mothers in particular far too often miss out on the emotional and financial support they require.

In one study, researchers interviewed more than 5,000 mothers from throughout the United States who were bringing their children either to a general clinic or an emergency department. Thirty-five percent of the mothers interviewed tested positive for depression, and, as it turned out, they had good reason to be sad. The depressed mothers were more likely to have lost federal aid, such as welfare or food stamps, and to consider their children in poor health. Patrick Casey, a medical researcher at the Arkansas Children's Hospital and a co-author of the study, warned that "policymakers who want to move families off welfare should consider evaluations for maternal depression if they want to optimize successful outcomes." Casey also suggested that being treated for depression should count towards welfare eligibility hours.

A mother's financial hardship can often be bad news for her child: Empirical studies have established a strong relationship between poverty and child abuse. Even in the rodent world, poverty and motherhood is clearly a dangerous mix: At the University of California at Irvine, Kristen Brunson, a postdoctoral fellow in anatomy and neurobiology, devised a rough model of low socioeconomic support for moms by removing bedding from a rat's cage at the time of delivery and replacing it with a paper towel. "It led the mother rats to become agitated and abusive toward their pups," Brunson says. "It really stresses a mother to take away the ability to give her pups a proper nest."

In northern California, Rick Hanson, a psychologist and the author of *Mother Nurture*, has proposed a new medical diagnosis applying to overstressed and undernurtured moms, whom he says can be literally depleted of important nutrients such as minerals and amino acids. He calls it Depleted Mother Syndrome and says it is so common that it warrants a new movement in support of mothers, on par with past activism for civil rights and the environment.

In contrast, as this book has described, mothers who feel they're enjoying sufficient support can be some of society's most motivated and productive members. In large part thanks to her secure situation at home, Jane Lubchenco helped inspire talented undergraduate students who might not have pursued a career in science for fear it would have prevented them from having a life with their families, says Frederick Horne, Lubchenco's former supervisor. He adds that Lubchenco's situation worked so well that he subsequently hired sixteen other faculty members in job-share arrangements, and that several of the mothers turned out to be award-winning teachers. "They are all better teachers than they might have been because they're so grateful and happy that they're being treated well and being respected and don't have to worry about their children," he says.

Lubchenco paid a price for her flexibility at work. Her family struggled on what amounted to a single salary until Horne helped get her and her husband a raise, recognizing that they were each working more than half time. "We did work hard, because we knew we were pioneers," Lubchenco says. "Sometimes we had to remind our colleagues that we weren't really each full-time." Yet Lubchenco says she and her husband feel "tremendously fortunate" to have had their opportunity, adding, "I think the reason we're both still going really strong is that we didn't get burned out trying to do the absolute impossible."

Catherine Gray, The Natural Step president, received similar extraordinary support from her board members after she found she was pregnant with Kai back in 2002, when her philanthropic business was still in a start-up mode. "When I found out I was pregnant I thought, oh my God, what does this mean for my career?" she recalls. "So I went to the board and said, 'Here's the situation. I know I will not be able to give the organization the kind of single focus I've given for the past five years.' And they turned out to be just incredible." The board members figured out a way to restructure the staff, giving two other officers more day-to-day responsibility. As it turns out, this allowed Gray to focus on what she does best: developing ideas, fund-raising, and dealing with the public. "It also allowed the organization to step up to the plate without me," she says. "And now we know they can. So what I thought would be a career-killing move turned out to be the best thing for my job. I pinched myself, thinking, why didn't I do this sooner?"

Women like Gray and Lubchenco have shown that, under the right conditions, parenting is not as draining and depleting as it's so often characterized, and that it can be a great source of energy. A Mommy Brain, as we've seen, can be an asset on the job and at home. As I describe in the next chapter, it can also be a potent force for social change.

CHAPTER
13

Political Drive

The Magic of Motivated Mothers

We are used to seeing what we call "a mother" completely wrapped up in her own pink bundle of fascinating babyhood, and taking but the faintest theoretic interest in anybody else's bundle, to say nothing of the common needs of all the bundles. But these women were working all together at the grandest of tasks—they were Making People—and they made them well.

CHARLOTTE PERKINS GILMAN,
HERLAND

CATHERINE GRAY WAS speaking to a large ballroom full of potential donors to her environmental think tank, The Natural Step, when she abruptly stopped talking about the group's recent achievements and started to address the future. Dressed with elegant severity—her long hair pulled back in a bun, a dark shawl over her black turtleneck—Gray gripped the podium with both hands. On a large screen behind her flashed a series of color portraits of the children of Natural Step employees, the last of which was Gray's own seven-month-old infant, David Kai.

"I stand here today as a *mom,* and what drives me now is a simple desire," she told the donors. "I want Kai and all the children you see up here to look at me in twenty years in disbelief. I want Kai to turn to me in twenty years and say, 'Mom, did you really pour *oil* to fuel your cars, polluting the air? Mom, did you guys really make bottles out of *plastic* that polluted natural systems, and use them once to drink, and then throw them in the garbage? Mom, did you really bury *nuclear waste* in the ground, and think that was okay? Did you guys really not get that we're all connected? Was most of your world really not thinking of these issues?'" By the time she wound up, Gray had completely assumed the future voice of her son. "Mom, thank you for standing up for what's right," she said. "Thank you for thinking of me and all the kids. Thank you for being part of the paradigm shift, and helping to turn things around."

Sitting in the audience that spring afternoon of 2003, I'd rarely heard anything so schmaltzy—or so touching. With a true Mommy Brain touch, Gray briefly burst into tears at one point, saying, "Blame it on the pregnancy hormones!" At the end, she brought Kai up on stage and told the donors: "There's someone here who wants to say thanks." Later, Gray told me her staff had encouraged her to follow her instinct to give Kai, and her own transformation, so public a role, even though, she said, "I'd felt totally out of my comfort zone" in doing so. As it happened, however, this philanthropic CEO was following a long and honorable tradition by which mothers make the personal political and, in doing so, extend a commitment to their own families' welfare to the world at large.

Contrary to the long-standing stereotype that women narrow their world view and passively retreat into the home once they have babies, it is just as often true that motherhood can energize and even radicalize them, especially when the cause is relevant to their children's welfare. Once engaged in the struggle of their choice, be it promoting safe bike routes to the local school, joining a national fund-raising drive for research on cystic fibrosis, or working for a presidential candidate, they can enlist capacities they've developed as mothers, such as perceptiveness, efficiency, resiliency, motivation and social skills, to great advantage.

In a heartfelt essay written when he was eighty-six years old, the pioneering neuroscientist Paul MacLean described this phenomenon in terms of brain anatomy. Based on his evolutionary model of the "triune" mammalian brain, MacLean has long proposed that the cingulate cortex, a part of the brain's limbic system and located in the middle of the cortex, was key in the development of the extensive maternal behavior that distinguishes mammals from other creatures. In his 1998 essay titled "Women: A More Balanced Brain?" he theorizes that in humans, maternal behavior has evolved to enlist the newer prefrontal cortex, allowing "a concern for the future welfare of the immediate family to generalize to other members of the species, a psychological development that amounts to an evolution from a sense of responsibility to what we call conscience." And he concludes that, because of this progressive brain development, "we can now say that for the first time in the known history of biology, we are witnessing the evolution of human beings with a concern not only for the suffering and dying of their own kind, but also for the suffering and dying of all living things."

Many of MacLean's ideas, including this one, remain controversial. And yet modern history provides plenty of examples of women who have channeled energies arising from their own domestic sense of responsibility into a much broader commitment for social reform.

Some Tactics of Militant Motherhood

Whether acting alone or as members of powerful groups such as the Mothers of the Plaza de Mayo or Mothers Against Drunk Drivers (MADD), mothers frequently exploit the very status that leads much of society to assume that their interests are narrowly focused on the home. In writing about health care for the *New York Times Magazine* in 2004, for instance, Senator Hillary Clinton noted that she had prepared her article at her kitchen table. Likewise, after the U.S. invasion of Afghanistan, the novelist Barbara Kingsolver complained, "I feel like I'm standing on a playground where the little boys are all screaming 'He started it!' and throwing rocks. . . . I keep looking around for somebody's mother to come on the scene saying, 'Boys! Boys!'"

Often, as Gray did, mothers have taken the more explicit step of literally holding up their babies, or images of them, as poignant bona fides. U.S. temperance activists—one of the most powerful women's movements of all time—published posters of mothers surrounded by children to stress how families were victimized by hard-drinking fathers. Even suffragettes printed pictures of children to suggest their maternal duties had earned them a political say. One postcard from those times shows a line of marching infants with a banner reading, "Votes for Our Mothers."

Perhaps the most important transformation involved in becoming a mother means becoming a protector equipped with a keen new awareness of the dangers in the world. As you've read earlier, and as Gray so explicitly acted out, this role comes with a heightened sense of empathy, an ability to see the dangers from the point of a potential victim. The consequent burden of responsibility so many mothers have felt has often tipped the scales between complacency and collective action.

The feminist leader Jane Addams recognized this potential in her 1915 essay exhorting women to seek the right to vote. Calling the ballot "an implement to preserve the home," she argued that if women were to fulfill their traditional duties properly, they had to "extend their sense of responsibility" and supervise those in charge of civic affairs: "If the garbage is not properly collected and destroyed, a tenement house mother may see her children sicken and die of diseases from which she alone is powerless to shield them, although her tenderness and devotion are unbounded," Addams noted, adding:

> She cannot even secure untainted meat for her household, she cannot provide fresh fruit, unless the meat has been inspected by city officials, and the decayed fruit, which is so often placed upon sale in the tenement districts, has been destroyed in the interests of public health. In short, if woman would keep on with her old business of caring for her house and rearing her children she will have to have some conscience in regard to public affairs lying quite outside of her immediate household. The individual conscience and devotion are no longer effective.

Imbued with this spirit, even decades before they won suffrage, millions of U.S. mothers waged grassroots campaigns through their membership in voluntary groups, which spearheaded important early social reforms. From 1830 to 1920, in the name of "social housekeeping," these associations, including the National Congress of Mothers (the precursor to the PTA) organized public lectures, letter-writing campaigns, and petition drives on behalf of such worthy causes as food and drug safety, pensions for widowed and destitute mothers (an effort that evolved into modern welfare programs), compulsory school attendance, child labor reform, public kindergartens, and free public libraries. "Many thought of themselves as mothers to the nation, as well as to specific children," says Theda Skocpol, a Harvard historian. "They believed that women had a moral sense that was stronger and in some ways more community-oriented than men's."

That era was a heyday for politically minded mothers, even as, or perhaps because of, the odd cultural contradiction of putting mothers on a pedestal while depriving them of any semblance of rights equal to men's. ("A woman is a nobody. A wife is everything," opined the *Philadelphia Public Ledger and Daily Transcript*, following the seminal Women's Rights Convention in Seneca Falls in 1848. "A pretty girl is equal to ten thousand men, and a mother is, next to God, all powerful.") The mid-nineteenth century temperance movement galvanized many women, notably including Susan B. Anthony, who never bore children, and the less-remembered but enormously influential Elizabeth Cady Stanton, the devoted mother of seven, whom she affectionately referred to as "miserable little underdeveloped vandals."

Both of these path-breaking women moved on to embrace the abolitionist movement, in which they gathered networking strength used for the next decades of battling for the vote. As a newlywed in 1840, Stanton traveled with her husband to attend the international Anti-Slavery Convention in London, only to find that the British & Foreign Anti-Slavery Society, which sponsored the event, had banned women from the convention floor. If they wished to hear the speeches, delivered by men, they'd have to sit in a separate, roped-off chamber. "It struck me as very remarkable," Stanton later recalled, "that abolitionists, who felt so keenly the wrongs of the slave, should be so oblivious to the equal wrongs of their own mothers, wives, and sisters."

Over the next many years, as Stanton and Anthony became fast friends, "Aunt Susan" would often help with the children, while Stanton continually drew on her inspirations as a mother to fuel her political efforts. Of an afternoon of collaborative multitasking, she once wrote: "We took turns on the domestic watch-towers, directing amusements, settling disputes, protecting the weak against the strong, and trying to secure equal rights to all to the home as well as the nation."

The Feisty Origins of Mother's Day

It's a relatively obscure fact that even the saccharine modern bonanza for the U.S. floral and greeting card industries we call Mother's Day originated during these early years when women claimed their political voice, and in an explicitly political context. Mothers had been celebrated since ancient times. The Greeks held festivals for Rhea, mother of the gods; the Romans honored Cybele, a mother goddess. The seventeenth-century British paid tribute to the Madonna, and, by extension, their own mothers, on Mothering Sunday, the fourth Sunday in Lent. But the U.S. holiday was initially intended less to honor mothers as such than to bring them together to work for change. During the Civil War years, a West Virginia teacher and homemaker named Anna Maria Reeves Jarvis organized Mothers' [*sic*] Work Days to improve local sanitary conditions, and, in the postwar years, help reconcile families whose sons had fought on opposite sides. In a separate effort, beginning in 1872, the suffragist Julia Ward Howe, a mother of six, and the author of "The Battle Hymn of the Republic," also lobbied for a national mothers' day, which she hoped would be dedicated to peace. After Jarvis's death in 1905, her daughter, also named Anna, pledged to carry on the tradition, doggedly campaigning to nationalize the holiday.

In 1914, President Woodrow Wilson acceded and declared Mother's Day a national holiday, to occur on the second Sunday in May. This new holiday placed more emphasis on a mother's family role than on her potential activism. Businesses soon appropriated the occasion, infuriating Jarvis's daughter. Towards the end of her life, she even filed a lawsuit to try to stop the celebrations and reportedly was once arrested for trying to disrupt a mothers' convention where women were selling white carnations to raise money. She protested that she had meant the holiday to be "a day of sentiment, not profit."

Still, whatever cheap bouquets and hollow tributes they've had to endure each spring, mothers have continued to follow their activist impulses when given sufficient incentive. Again and again, they've exploited society's clichés about them for their own ends. On various occasions, for instance, mothers have used the image of the nonconfrontational mom as a weapon to combat war. During the Vietnam War years, women in Southern California joined forces as the benevolently dogged Another Mother for Peace, which soon became a prominent national movement. "Our appeal was that as a mother, you bring these wonderful people into the world, and you really feel you must make it a good place for them to live," Gerta Katz, one of the founders, told me in 2003. At the time, Katz, then well into her seventies, was helping the group revive its operations to oppose the U.S. war in Iraq.

Like the *madres* of Argentina, mothers quite often are at least initially moved to organize only after they've suffered a personal loss. It's almost as if the new reserves of mental capacity and drive summoned to care for a new child are so powerful that they must head someplace else, once that child is gone or compromised. So it was in 1979, when Candace Lightner, whose thirteen-year-old daughter was killed by a repeat drunk driver, and Cindi Lamb, whose five-month-old daughter became one of America's youngest quadriplegics in a similar accident, formed MADD, which soon claimed chapters throughout the United States, and eventually also in Canada, Australia, Sweden, and Japan. Similarly, concern for her young children's recurring health problems spurred Lois Gibbs, an upstate New York homemaker, to organize others in her neighborhood, known as Love Canal, after she learned that her home was sitting atop 21,000 tons of buried chemical waste. Gibbs subsequently became known as the Mother of Superfund Legislation, the federal U.S. laws she inspired, which regulated cleanups of other massive toxic dump sites.

Still, although Lightner, Lamb, and Gibbs "extended their responsibility," as Jane Addams would have put it, only after they had painfully personal reasons to do so, other mothers have been drawn together by the sheer increased anxiety about the world in which they were bringing up their children. So it is with Mothers Acting Up, launched in May of 2002 in Boulder, Colorado, which encourages its members to contact their elected officials on issues such as global warming, genetically modified foods, and spending on education. As if echoing Addams, C. C. Pelmas, the mother of two boys and one of the group's leaders, says she had increasingly felt challenged by the irony of worrying about such things as healthy lunches and a safe neighborhood, "while letting the larger hopes and desires (for a healthy planet, a world full of joy, and a knowledge that all children around the world are being well-taken care of) lie deep inside me with no voice or validation. I don't need to speak to every mother to know that these desires are universal."

This sentiment once again squares with Sara Ruddick's view of "maternal thinking" as a distinct world view, and set of values, arising from the daily work of caregiving and what she calls "preservative love." According to her hopeful argument, the practice can amount to repetitive submission to Gandhian principles of nonviolence such as reconciliation and resistance to injustice. As late as 2001, in an essay written just a few days after the tragedy of September 11, 2001, Ruddick reaffirmed her hope that "people who make the work of caring for children an ongoing and serious part of their working lives may acquire ways of thinking and acting that help to create and sustain a culture of peace.'" Even so, in an interview in 2004, Ruddick sadly conceded that there has been little proof that parenting, in itself, makes mothers or fathers more peace-loving.

At times, it almost seems like the reverse is true. Consider the warlike mothers of the ancient city-state of Sparta. "Return *with* your shield or *on* it," is how these matrons bid their sons farewell, as they went off to battle. That's because the soldier's heavy, cumbersome shield was the first thing he'd cast aside in a retreat, yet could also serve as a stretcher. Mothers, alas, have also been active in right-wing hate movements from Nazi Germany to the U.S. Ku Klux Klan and urban teenaged skinheads. The historian Claudia Koonz, author of *Mothers in the Fatherland: Women in Nazi Germany*, suggests this phenomenon may be a product of women's historic acquiescence to men, nostalgia for traditional roles and longing for stronger communities, all of which can be aggravated in times of economic crises and fear. As Virgina Woolf has written, "Those who are economically dependent have strong reasons for fear."

Were such factors at play in the 2004 U.S. election, when, according to one polling firm, a majority of women (51 percent) voted for the Democratic Party challenger Senator John Kerry, but an even stronger majority of married mothers (56 percent) voted for the incumbent, George W. Bush? Intense speculation focused, during the campaign, on the potential influence of so-called Security Moms, who, pundits said, were seeking a decisive leader in the new era of terrorism fears. Bush's strategist, Karl Rove, was reported to believe that a shift among women with children under 18 was a major factor in the Republican Party's historic triumph in the 2002 midterm elections, in which Bush became the first Republican in a century to see his party gain seats in an off-year election. "Since 9/11," said Debbie Creighton, a thirty-four-year-old Santee, California, mother of two who voted for Bill Clinton twice before leaning to ward Bush, "all I want in a president is a person who is strong."

Although women at large have at least since the 1980s favored Democrats—accounting for a famous "Gender Gap"—married mothers have long been more conservative. (They are also more reliable voters, registering and turning out in greater proportions than single women.) "What drives them are their children," says public opinion consultant Ethel Klein, who organized focus group research on mothers in those years. This rang true especially in 2004, when a group of voters were asked if they feared a member of their family would be victimized in another terrorist attack. Just 17 percent of men, but 43 percent of women *and 53 percent of mothers with children under 18*—said yes. Critics of Bush, who question his intelligence, may find U.S. mothers' loyalty to him reason to wonder about this book's premise, that motherhood can make women smarter. Yet, as we've seen, the motivation to keep children safe is one of the most powerful forces affecting mammal mothers, and in many of us it may simply overwhelm more sophisticated analyses.

On the other hand, history is full of examples of mothers who have thought for themselves and bravely confronted the status quo. One of the boldest of these, once hailed as "the most dangerous woman in America," was a white-haired, lacey-bloused firebrand of the nineteenth-century labor movement, who called herself "Mother Jones."

How Mother Jones Found a New Family

Born in Cork, Ireland, in 1837, Mary Jones was thirty years old, living in Memphis, and engaged full time in the care of her husband and four children, when she lost them all to an epidemic of yellow fever. Her youngest was then not yet a year old. She washed the bodies for burial and sat grieving, alone. No one came to her aid, as she later wrote, because all around her, other families had been hit just as hard.

Virtually overnight, Jones was transformed from an overworked soul in constant service to five other people to a woman who had to care only for herself, yet who also had to create a new life's mission from scratch. Eventually she moved to Chicago, where she entered the dressmaking business and worked for "aristocrats" who lived in luxury in mansions on Lake Shore Drive. She was struck by the contrast between these "lords and barons" and the "shivering wretches, jobless and hungry, walking along the frozen lake front." Such scenes inspired her, within a few years, to become active in the Knights of Labor, which was gaining force in the 1880s. From there, it was a short path to her mission, and her fame.

Soon, Mary Jones's flaming speeches had become part of the sensational street theater of strikes at coal mines, steel and textile factories, breweries and railroads. In motherly defense of her downtrodden workers, she strategically took on the larger-than-life persona of Mother Jones, exaggerating her age, wearing old-fashioned black dresses, and alluding often to her impending demise. Writes her biographer, Elliott Gorn: "By 1900, she had stopped referring to herself as Mary altogether and signed all of her letters 'Mother.'" In time, laborers, union officials, even presidents of the United States addressed her that way, and they became her "boys." Gorn notes that the Mother Jones persona served Mary Jones as much as it served her cause:

> It freed her at a time when most American women were expected to lead quiet, homebound lives. Ironically, by making herself into the symbolic mother of the downtrodden, Mary Jones was able to go where she pleased and speak out on any issue that moved her. She defied social conventions and shattered the limits that confined her by embracing the very role that restricted most women. . . . She ignored family life and lived entirely in the public realm, or more precisely, widened the family circle to embrace the entire family of labor.

As Mary Jones so well understood, her title's hefty emotional punch lay in its summoning up of the associations most of us have with a (supposedly) selfless guardian: one who watched out for our safety, kept us warm, taught us right from wrong, and cleaned up our messes. Time after time, politicized mothers like Kingsolver and Clinton have played on these themes, offering images from the nursery or metaphors cooked up in the kitchen. The Argentine *madres'* trademark is the white scarves they wore on their heads, originally made from the swaddling cloths of their disappeared children and lovingly embroidered with their names. Code Pink, a women's activist group that holds vigils in front of the White House to protest war and environmental abuse, derives its name from the alarm used in hospitals to warn when a baby has been abducted. And on the Web site of the Million Mom March, a 1999 movement against gun violence, you can click on icons to find public figures deserving of either the "Time-Out Chair" (when I last clicked, it was John Ashcroft) or "Apple Pie" (Michael Moore).

Back in 1984, when congressional Democrats were desperately searching for a sound bite to challenge the genial President Ronald Reagan, Pat Schroeder (D-Colorado), found inspiration while scrambling eggs one morning for her children. Reagan thereafter became known as the "Teflon President." Boasting a bit, mothers in public roles often portray themselves as efficiency experts—or are portrayed as such by others. When Governor Jennifer Granholm of Michigan, a rising Democratic party star and the mother of three, was asked about the statehouse mess she had inherited, she replied: "Like every good mother, I'm used to cleaning up other people's messes."

This was also Patty Murray's image, years before she became the first U.S. senator with children still at home, in 1992. Murray debuted in politics by leading a grassroots coalition of 13,000 parents to save a local preschool program from budget cuts. She went on to serve on the local school board, and in the Washington State Senate, before being elected to the U.S. Congress and then to the U.S. Senate. There, she gained fame as "the mom in tennis shoes." The *Seattle Post-Intelligencer* called her "a workhorse, not a show horse."

As Granholm and Murray demonstrate, mothers have gradually but relentlessly moved into prominent roles in U.S. politics, where they can apply "maternal thinking" not as outsiders but as an emerging part of the Establishment. In the first years of the new millennium, there were more mothers than ever before in the U.S. Congress and Senate, just as they had increased their ranks in prestigious fields such as law, business, and science.

Change Motherhood and Change the World

The growing numbers of mothers throughout the workforce, a trend that reached historic levels by the turn of the twenty-first century, has been accompanied by an increasingly powerful dissatisfaction. Although some upper-class women may have fled the corporate rat race to channel their competitive instincts into local school boards and soccer teams, those who can't afford that retreat have been frustrated by what seems to be their sole option of shunting their babies off to daycare by six weeks and relegating "quality time" to the drive-through line at the McDonald's. That kind of pressure leaves parents and children feeling exhausted, and lonely, far short of the cerebrally enriched ideal. It also leaves them often furious with a culture that, while constantly promulgating all the newest theories of how much children need strong attachments (to moms, dads, or hard-to-find loyal caregivers) offers less than the minimum support necessary to provide them.

So what's to be done?

Some solutions may be brewing from various factions of a new reform movement that has been gathering force in the early years of the millennium, led by commentators and advocacy groups demanding more recognition and support for the basic job of childcare. At this writing, the movement is heading along two separate tracks. One faction seems guided by the "maternalist" spirit of the nineteenth-century women's reforms, with its ostensible focus on the welfare of children. The other campaigns for a better deal for mothers themselves, and it goes beyond the mere symbolic recognition of the value of their work to encompass direct economic support.

The first group is spearheaded by the Motherhood Project, launched in 2001, and led by Enola Aird, a Panamanian-born graduate of Yale Law School, who has frequently called on her own experience in her advocacy of more "balanced" lives. Aird is a model foot-soldier of the opt-out revolution. She spent the first seven years after her daughter was born pursuing a high-powered and high-paying law career. But after a day when guilt caught up with her—her daughter was ill, but she'd chosen to leave her with the sitter—Aird quit the job, pronouncing herself "totally exhausted and terribly stressed out."

Back at home, Aird quickly filled up her hours once again plunging into work for local, then state-wide and national family and human services organizations, including a stint with the Children's Defense Fund in Washington, D.C. Meanwhile, she gave birth to another baby and took on the care of her elderly parents and aunt. Showcasing her fervent effort to fulfill all these obligations, as well as her new advocacy role, Aird scheduled our interview, in late 2003, for the half-hour she was driving her elderly parents on errands. Reality fell short of the ideal, however, as the static on her cell phone headphones and background noise of the elderly relatives chatting in the back seat made us repeat ourselves more than a few times. And her battery was about to wear out. When I said I was writing a book suggesting ways in which motherhood might make you smarter, her initial response was, "You're joking, right?" Yet, the next minute, she said that motherhood and mothering work had changed her dramatically, leading her from thinking narrowly about her own two children to thinking about our culture. "Mothering does all kinds of things to your mind, to your spirit and how you feel about yourself. But I don't know if it does the same in a culture that doesn't support you as in one that does," Aird says.

Clearly, for too many women, it doesn't. But did Aird know how to fix that? When we spoke, she said she was still trying to find out, in part by overseeing a national study, to be released in 2005, receiving the opinions of 2,000 mothers. She said she hoped the findings would produce specific issues to work on and give clues to possible solutions in the context of a "mothers' renaissance" that would emphasize such "nonmonetary values" as "caring and connectedness."

The group had also staked out the issue of television advertising, which Aird believes is part of a culture that is "toxic to children." The project kicked off its efforts with a warning to advertisers that mothers involved in the project would oppose both marketing and marketing research in schools, in addition to the targeting of ads to children younger than eight. Aird has briefed Congress on the impact of ads on children, and she and her appointed "Mother's Council" have written a letter to the chairman of CBS, Leslie Moonves, protesting the ads on the 2004 Super Bowl.

The Motherhood Project's strong advocacy of the value of stable marriages and of mothers being the primary source of caregiving in a family has led some critics to worry that the group's agenda is to keep moms at home. Fueling this concern is that the project is sponsored by the conservative Institute for American Values, whose other associates (including author Maggie Gallagher, author of *The Case for Marriage*), have made comments appearing to endorse the old-fashioned stay-at-home-mom/breadwinner-dad model. Aird says she doesn't share these views, and supports women who choose to work and depend on other caregivers. Yet the project's Web site declares that "mothers are central to children's development; we are not easily interchangeable with other people."

The second mothers' faction has been going about its job differently, focused as it is on mothers' needs for economic and social support, independent of their children. Joanne Brundage, founder of the national advocacy group Mothers & More, says this has created an organizing hurdle. "Motherhood brings out the lioness in women, but what we've never seen until recently is moms getting organized on their own behalf. It's a bit at odds with our nurturer image," she told me. "Still, we're being harder-edged about it. We're an organization that is here first and foremost to encourage the nonmother pieces to flourish while you're being a mother. Our feeling is that this is the unfinished work of feminism."

Brundage was a thirty-five-year-old letter carrier in 1987, when she decided to stay home after the birth of her second child. "We were braced for the financial hit, but I was knocked off my feet by the loss of my identity," she says. "I signed the resignation papers, put my three-month-old in his Rock-n-Ride, and drove around sobbing for the next two hours." She put an ad in her local newspaper in Elmhurst, Illinois, and after hearing from several like-minded mothers, formed a group originally called F.E.M.A.L.E., for Formerly Employed Mothers At Loose Ends. Its more than 7,000 members in 180 chapters do grassroots work as simple as "baking that casserole when you're having your baby," or as potentially far-reaching as lobbying state officials for new benefits for caregivers.

Brundage says she still sees a great need for "consciousness-raising" among her own group's diverse members, many of whom, she says, just aren't ready to go out and demand more fairness. She therefore supports the goals of a harder-hitting advocacy organization called Mothers Ought to Have Equal Rights, inspired by Ann Crittenden's landmark 2002 book, *The Price of Motherhood,* which takes on workplace rules, marriage laws, government codes, and cultural conventions that, as the author argues, combine to make motherhood the world's least valued job. Although most barriers have been removed for "unencumbered" women who can act like men on the job, working overtime and traveling without a hitch, Crittenden says, they pop up as never before if these women choose to start a family, with repercussions that can last a lifetime. Mothers have smaller pensions than men or childless women, and American women older than sixty-five are more than twice as likely to be poor as men of the same age.

To address these concerns, Mothers Ought to Have Equal Rights, which Crittenden founded with another feminist author, Naomi Wolf (*The Beauty Myth* and *Misconceptions*), has drawn up an Economic Empowerment Agenda geared toward establishing in the United States benefits similar to those enjoyed by caregivers in Britain, Canada, France, Belgium, Holland, and Scandinavia. Included on the far-reaching wish list is paid family leave for each parent, a refundable tax credit for anyone caring for a family dependent, social security credits for unpaid family caregivers, living wages and better training for paid caregivers, guaranteed flextime for parents of infants and toddlers, and the inclusion of unpaid labor in the home in the national Gross Domestic Product.

By the first years of the new millennium, the growing attention on the varied plights of children and caregivers had begun to catch the attention of U.S. politicians. Activists in more than twenty states were seeking legislation similar to California's paid family leave law. Mothers' advocacy groups also hailed the start, in January 2003, of the new national Time Use Survey by the U.S. Census Bureau, which, among other things, has offered evidence, cited earlier in this book, of how many hours U.S. mothers and fathers are dedicating toward the care of young children. Some saw this as a key first step towards making a case for social security benefits for unpaid caretakers.

Still, the mothers' movement clearly had a long way to go before its achievements could start to catch up with its rhetoric's promise. One difficulty drawing attention from critics was that its most vocal faction was also its most privileged: the women who had the comparable luxury of being able to stop work, even temporarily, to dedicate themselves to parenting, and who were calling for social security credits. Especially in a time of war and nationwide economic hardships, this risked an appearance of elitism, a charge that had been leveled at the feminist movement of the 1970s as well. But a much more central problem was the still too-slow progress in engaging fathers to shoulder more duties at home. Caring for dependents remained, for the most part, an unpaid job, an invisible job: a woman's job.

Changing this status quo would free mothers to use more of their smarts and drive outside their home, for the sake of their children, and others'. But it would take a domestic transformation, as well as an upheaval at most workplaces, which now as a matter of course place the greatest store on the most unencumbered employees. Nearly a century ago, the feminist-socialist writer Charlotte Perkins Gilman offered a fictitous vision of a society that might solve these problems. "Herland" was set in a tropical land that had been walled off by an earthquake, after most of the men were killed in battle. Isolated from the rest of the world for 2,000 years, the women had taken over, and in time, somehow, achieved the power to have children—all girls—by themselves.

Being naturally oriented toward creating quality rather than quantity, this country of intelligent mothers had carefully planned its population to maximize resources, prevent poverty, and encourage education and contentment. Gilman tells the story through the eyes of three male explorers who stumble upon the land, and are held there first as prisoners, then as guests. Yet once the women learn the state of the world outside their borders, they insist the men leave without revealing to anyone else what they've discovered. The men are eventually persuaded to agree.

Gilman's book is clearly unfair to many good men, as well as sanguine to the point of being silly about women. Still, one compelling truth shines through in her story, as it does in Sara Ruddick's *Maternal Thinking*, and I hope also in this book. It's that the simplest ethic of motherhood—care for others—can be a potent way of being smart. In the words of Paul MacLean: "Why else are we here, if not to help one another?"

CHAPTER

14

Neuroscientists Know Best

Ten Tips to Help You Make the Most of Your Mommy Brain

MANY WOMEN GO through motherhood without appreciating its transformational potential—"almost as if they're driving around with a treasure hidden in the trunk of their car," says neuroscientist Kelly Lambert. If you've read this far, you've already taken the first step in putting that treasure to use. Conscious of yourself as profoundly changed by motherhood, you may be more sensitive to what you're experiencing and perhaps also more deliberate in how you react. Psychologists call this "metacognition," meaning that you're thinking about your thinking. In earlier chapters, I've quoted scientists' advice about how women can become smarter through this new awareness. Now, here are some specific suggestions to help you along:

Don't "Surrender to Motherhood": Take Back Motherhood

Remember the Australian neuroscientist Allan Snyder's observation that new mothers are like Albert Einstein, with much of their brains focused on matters of supreme importance. Keep in mind that a Mommy Brain—in some ways, just like a baby brain—is a super-attentive learning machine, engaged in life-or-death work, and in training for skills that can be used for years to come.

While you focus most of your energy on your child, you might also explore your own new potential, assuming you have energy to spare.

"This may be the time to try something very new for the brain," says Lambert. During one of her own maternity leaves, after she had observed the remarkable growth of new brain cells in her lab rat moms and realized just how dynamic the maternal brain can be, she decided to take advantage of what she believes is a "window of plasticity." Says Lambert: "After being a science nerd for too many years, I was becoming concerned that the areas of my brain involved in more artistic endeavors were probably filled with cobwebs." So while her baby napped, she took out her paints. "It actually seemed like I had more 'ability' than on previous attempts," she says, adding: "Everyone I knew got a painted cake plate that Christmas."

Don't fall into the trap of thinking of your Mommy Brain as an impaired brain. Recognize that as much as you may tell yourself you don't really believe mothers are mentally handicapped, you may, deep down, harbor a bias that can set you up to fail. "This is what older adults do," says Julie Suhr, the Ohio University neuropsychologist. "They say they don't believe in the stereotype that older people are mentally impaired, but then tests show that they actually do. You incorporate cultural expectations; they suck you in, and become part of your identity." Suhr offers new mothers the same advice she gives aging adults. "Remind yourself that the stereotype exists. Remind yourself that everyone is forgetful and distracted at times, rather than attributing it to some permanent or semi-permanent problem. And then externalize it: what's happening *outside* you that might be getting in the way?"

Recognize Your New Priorities

A "very strong attentional focus" is one of the "unequivocal neurological advantages" accompanying a Mommy Brain, says brain plasticity expert Michael Merzenich. Still, it's important to keep in mind that the first place this new focus will and should usually be trained is on your baby. There's no reason you shouldn't be able to marshal some of your new powers of attention when working outside the home, Merzenich says. But you'll probably have to make a more conscious effort to keep your mind on task. The first way you can prepare is to make sure your baby is safe and well-cared-for while you're at work. If you can afford it, you might also try to find ways to lessen your workload, particularly during the baby's first three years, so that you can give yourself more time to concentrate on mothering, and so enjoy more peace of mind.

Remember that starting off on the right track with your baby, with a lot of love and attention, will give both of you the best chance of success for many challenging years to come. The behavioral neuroscientist Alison Fleming notes that women who have positive interaction with their infants early on tend to have better self-esteem. But there's still hope for women whose early experience isn't ideal. Ravenna Helson, the UC Berkeley psychologist, says she has found that mothers who sought counseling in parenting skills could improve both their relationships with their children and their own feelings of competency.

While at work, try employing some practical tricks to keep yourself focused. Edward Hallowell and John Ratey, both psychiatrists and the authors of *Driven to Distraction: Recognizing and Coping with Attention Deficit Disorder from Childhood Through Adulthood*, offer advice that might be just as useful for besieged working mothers as for adults with attention deficit disorder. They advise breaking down large tasks into small ones to avoid feeling overwhelmed, and setting deadlines for yourself for the small parts. They also recommend following the OHIO (only handle it once) rule when it comes to paperwork. If you're going to respond at all to a bill, letter, or memo, try to do it right away instead of letting your to-do piles reach depressing heights. (One of my own tricks is to maintain an oversized calendar. When I find myself obsessing about something that doesn't need to be done right away, I write a to-do note reminding myself to worry about it at some future time and then forget about it until then.)

Don't Underestimate the Power of Sleep

If there is any aspect of a Mommy Brain that's scientifically established to mess with your mind, it's lack of sleep, and, alas, a serious sleep deficit is all but unavoidable in the first weeks and months of parenthood. That's why my favorite advice among the millions of words that have been written on this topic is to share the load with your significant other. This issue will very likely be the first big test of your parenting partnership. If you get off to a good start, it's a wonderful omen for the next eighteen years. Shouldering the sleepless burden equally is a powerful demonstration of your commitment to each other, and it also helps both of you bond with your baby.

In this effort, it's best to set up clear rules from the beginning, and stick to them. James Maas, the Cornell University sleep expert, advises parents to trade off with three-night shifts, a tactic that keeps each partner as well rested as possible during the hazing period of early parenthood. This seems a lot smarter than one common recommendation that has Dad diaper the baby and then hand him to Mom, which means both of you end up exhausted. Breastfeeding moms should feel free to leave a bottle by the bedside. (Our pediatrician in Brazil had a useful idea here: Mix a spoonful of rice cereal in with the milk so that the baby feels fuller, and, if you're lucky, sleeps longer.)

Once your child is old enough to understand a spoken warning, you might want to adopt one of my favorite Family Rules, courtesy of the writer Barbara Ehrenreich: "Never wake a sleeping adult." Let your child know that your sleep is fundamental to the happiness of both of you. And don't sabotage your z's in other ways, advises Maas, who warns new parents to stick to a regular sleep schedule as much as they can, not to drink caffeine after 3:00 P.M., and to avoid alcohol within three hours of bedtime.

Improve Your Spin Control

If you take home just one lesson from this book, I'd recommend memorizing the mantra "This is a learning opportunity." Rather than think of your brain as being stressed, imagine it stimulated. Has your five-year-old thrown himself screaming on the floor in the video store because you won't rent *Halloween 7*? Try not to spend all your mental energy wishing that you could be beamed out of there. Reframe the situation as a challenge, and re-imagine yourself as its potential beneficiary instead of victim. How spontaneous can you be? Instead of something as predictable as whispering that he will never, ever, see his Game Boy again, use your Theory of Mind. What does your child really want? To show he's boss, of course, counting on your being pressed for time and worried about dirty looks from strangers. Try blowing his little mind by sitting down beside him, waving to the next people in line in front of you, and telling him calmly that you're just going to wait until he's done with his tantrum. Remember, while you're doing so that metacognition in such situations can not only relieve your stress but also teach you to deal more skillfully with adults who also throw tantrums.

Engage Oxytocin

As the Swedish neuroendocrinologist Kerstin Uvnas-Moberg notes in Chapter 6, recent findings about the potential influence on the brain of the peptide hormone oxytocin should inform new mothers' choices, even before delivery day. Oxytocin, as explained in Chapter 6, has been linked with memory and learning, in addition to maternal-infant bonding and good-for-your-brain social ties. Vaginal as opposed to caesarian deliveries and breastfeeding boost your exposure to this hormone, as might massages and skin-to-skin cuddling. But don't confuse natural oxytocin with pitocin, the synthetic oxytocin frequently used to speed up labor at U.S. hospitals. As scientists learn more about oxytocin's impacts on the brains of a mother and her baby, some have grown concerned that administering the synthetic hormone could have unintended consequences, even though doctors have assumed that oxytocin in the blood won't cross into the brain. "It could impact the system in ways we don't clearly understand," says Diane Witt, a psychologist and the program director for behavioral neuroscience at the National Science Foundation. "Babies' brains could be impacted at a critical time, which could cause learning disorders."

Socialize

Motherhood can yank the most determined misanthrope into new ways of relating to people and a new array of social contacts. But it can also isolate you, if you're stuck in the house most of the day with just your baby. Serious postpartum depression, which strikes one in ten mothers, or just a mild case of the blues, can be aggravated by loneliness. But regular and supportive social connections can do wonders for you and your baby.

Strong proof of this power of social support has come through in a twenty-five-year controlled study of several hundred low-income mothers in several U.S. cities, led by David Olds, a pediatrics professor at the University of Colorado. In Olds's project, one randomly assigned group of mothers was visited by nurses just once a week during their pregnancies until six weeks after their babies were born, the visits tapering off through the child's second year. During the ninety-minute meeting, nurses kept their attention determinedly more on the mother than the baby, discussing how she felt, how much family and friends were helping her, and what the mothers might do to improve their relationships. Follow-up research found these visits led to dramatically better lives for the children, who as a group had fewer injuries associated with child abuse and neglect, fewer teen pregnancies, and less dependence on government welfare.

The modern notion that mothers can be solely responsible for babies is of relatively recent origin and contrasts with a need for social networks that may be rooted in our genes. The anthropologist Sarah Hrdy says it's likely that humans evolved as "cooperative breeders," noting that in surviving traditional cultures, such as the Efe and Aka pygmies of Central Africa, babies can have as many as about a dozen different caregivers, allomothers, who are in contact with infants about 40 percent of the time.

Strollercize

Although working out may often seem like a luxury, it's actually a brain maintenance strategy that will help you feel and think better, which is usually better for everyone who depends on you. Physical exercise increases circulation to the brain, and as Harvard's John Ratey notes, it can even increase the number and density of blood vessels in the motor cortex and cerebellum. The three major neurotransmitters involved in mood, cognition, and behavior are all enhanced by exercise, Ratey says, and many studies have shown that vigorous physical activity helps ward off depression. Recent research also suggests that exercise involving learning complex movements, such as dance steps or yoga, can help sharpen memory and learning capacity in general. If you can't persuade your partner or a sitter to free you to go to the gym, there's a whole industry of prenatal Yoga teachers, Mommy & Me classes and informal jogger-stroller groups more than willing to help you out. Exercising in groups is also a great way to meet other mothers.

Mother Thyself

Laura Lopez, the mother of three and a frequent flier in her job as director of a United Nations publishing group, compares caring for children to preparing for crash landings in airplanes. "The flight attendants always tell you to put on your own oxygen mask before you help your child," she says. "If you're gasping for air, you're not going to be of much use to anyone else."

This advice can translate to finding ways to take even brief breaks from work and motherhood responsibilities as little gifts to yourself that can also be mentally refreshing. It can be anything from allowing yourself to have an uninterrupted conversation with a friend, despite how much strategizing that can take, to sneaking out of work early to see a movie. A friend of mine who frequently travels out of town has confessed that, when telling her husband what time she'll be back, she occasionally tacks on an extra hour or two, so that she can go out for coffee without feeling she owes an explanation. On the other hand, and while I'm not the first to suggest this, it's also a good idea to find time to be with your partner as a twosome to remind each other why you had children in the first place. As Lambert cautions, it's important to remember that acquiring a Mommy Brain doesn't mean you've lost your "wife brain," "daughter brain," "friend brain," or "professional brain." "You've just added an important wing to the brain's foundation, not torn down the existing structure," she says. "Being able to take the Mommy Brain hat off—even periodically, by focusing attention on other aspects of our lives—forces us to reassess and refocus when we return to the children, sometimes realizing toxic patterns that are developing, and redirecting ourselves."

While mothering yourself in this way, don't forget to take your own mother's most frequent advice, and watch your diet. A balanced diet is demonstrably good for the brain. Studies have found that antioxidants in fruits and vegetables can help prevent declines in brain function due to aging, and that leafy green and cruciferous vegetables (including broccoli, cauliflower, romaine lettuce, and spinach) are particularly helpful for older women's memories. And fish really does seem to be "brain food," as your mother may have called it; some research suggests that eating fish once a week could greatly reduce your risk of Alzheimer's disease.

Multitask Away—Within Limits

As you acquire mastery in multitasking, you may be tempted to do it all the time, piling on extra challenges like a circus clown juggling on a tightrope. There's good reason that multitasking normally works well for mothers: David Meyer, a University of Michigan cognition expert, has conducted research showing that people can, through training, improve their capacity to do more than one thing at a time. Yet Meyer cautions that people tend to overestimate their multitasking skills, and that there's a limit to even the most determined multitasker's ability.

Meyer and other researchers have found that switching between complex tasks can slow down an average adult by up to 50 percent. Intense multitasking can also cause stress that, when prolonged, can interfere with short-term memory. Of course, resisting contemporary pressures to multitask may seem tantamount to cultural rebellion. But before you cross over the line to inefficiency and burnout, remember Kathy Mayer, who, when her twins were breastfeeding, multitasked all day, pumping milk while driving, talking on her cell phone, and supervising doctors all over Denver. Mayer consciously relished the opportunity to come home and do just one thing: hang out with her children. One of the best things for your brain about multitasking can be how good it feels when it stops.

Change the World—Starting with Your World

Ravenna Helson says she has seen from her study of Mills College graduates that a mother's emotional and mental development depends to a great degree on the context in which she mothers. Mothers themselves can often determine a great deal of that context. If you plan to have children, make careful decisions when choosing a spouse—and a boss. Once you've got the kids, encourage them to help their dads with the chores. This lessens your own workload and stress while you're training a more considerate future generation. If your kids balk, cite research indicating that children who do housework with their fathers are more likely to have more friends. If your husband balks, cite other research that shows moms are more likely to find their husbands sexy if they help out at home. Pay attention to the news, especially political developments concerning government support for family leaves and daycare, and if you don't like it, as the legendary San Francisco Bay area newscaster Scoop Nitzger used to say, go out and make some of your own.

ACKNOWLEDGMENTS

Nothing is more humbling than rearing children, but writing a book comes close. I am indebted to scores of people who generously took the time to help me with this project.

This includes every one of the scientists and mothers quoted in these pages. But I owe by far the most to seven brainy moms: Kelly Lambert, who nurtured the book from the start, and whose insights sparkle throughout; Michelle Tessler, who skillfully helped shape the proposal; Amanda Cook, the brilliant co-architect of the book's structure; Jo Ann Miller, who made many very wise improvements; Elizabeth Share, who deserves credit for the title, numerous readings, and a treasured friendship; and my great friend and beloved sister, Jean Milofsky. This book is dedicated with much love and admiration to my own mother, Bernice Ellison.

Six smart men also provided vital insights and support: Craig H. Kinsley, Jeffrey Lorberbaum, and Michael Merzenich; and my father, Ellis, and brothers, David and Jim.

Marian Diamond, Sarah B. Hrdy, and Sara Ruddick not only provided me with inspiration fundamental to this book but graciously found time to discuss their ideas. Special thanks also to Bob Bridges, C. Sue Carter, Anne Gelbspan, Alison Fleming, Ravenna Helson, Marco Iacoboni, James Leckman, Michael Numan, Kevin Ochsner, Karen Parker, Stephanie Preston, Robert Sapolsky, Rhonda Staudt, and Kerstin Uvnas-Moberg for so graciously responding to more calls and emails than their schedules legitimately allowed. I am likewise seriously indebted to Jennifer Blakebrough-Raeburn, Ellen Garrison, and Kay Mariea for their professional skill, and, above all, patience. Friends who shared oxytocin and laughter include Nancy Boughey, Michelle Bullard, Gretchen Daily, Nancy Efsaif, Emily Goldfarb and Norma del Rio, Laura Hamlin, Laura Lopez, Maria Newman, and Idie and David Weinsoff.

Finally, this book obviously couldn't have been written without my husband and best friend, Jack, and our sons, Joey and Joshua.

NOTES

Unless otherwise indicated, all direct quotations are drawn from my interviews with those quoted. I conducted interviews by telephone, e-mail, and in person.

Chapter 1: Smarter Than We Think

3 *smart: Merriam Webster's Collegiate Dictionary*, 1108.

3 *If you've left the crayons:* Carlotta Eike Stankiewicz, http://www.people.virginia.edu/~pna5a/mommy.html.

4 *When researchers showed audiences*: Cuddy and Fiske.

4 *Today, nearly three-fourths of mothers with children aged one or older:* U.S. Census Bureau, "Fertility of American Women: June 2002," Oct. 2003.

4 *In* The Feminine Mystique: Friedan, 305.

5 *Friedan's brain-dead momma:* Kaufman.

5 *The doom-saying didn't end:* Waldman, 11.

5 *"It was as though my ovaries":* Quindlen, "Flown Away, Left Behind," http://msnbc.msn.com/id/3868018/.

5 *Polls in recent decades:* Stearns, 1.

8 *I was encouraged early on by a report:* Michelle Pridmore-Brown, "Breed Old, Die Late and Leave a Beautiful Brain," *Salon.com*, March 26, 1999; www.salon.com/mwt/feature/1999/03/cov_31featurea.html. See also Kinsley 1999.

8 *Including one headline that boldly declared: "Motherhood Makes You Smarter":* Maggie Fox, "Motherhood Makes Women Smarter, Study Suggests" (Reuters, Nov. 8, 1999).

9 *In February 2003:* Ellison, 46.

11 *The average contemporary working woman was spending about twice:* "Study Confirms It: Women Outjuggle Men," *New York Times,* Sept. 15, 2004.

Chapter 2: "Honey, The Kids Shrunk My Brain!"

13 *"Whatever does not kill":* Nietzsche, 5.

13 *"After an hour"*: Peri and Moses, 12.

14 *"And even "pregnancy dementia"*: Brett and Baxendale, 339–362.

14 *"Profusely blinde . . . ":* This is from John Donne's "The First Anniversary. An Anatomy of the World." The poem can be found at http://darkwing.uoregon.edu/%7Erbear/donne1.

14 *In 2001, two British neuroscientists*: Brett and Baxendale, 339–362.
14 *Although two small but well-publicized*: Buckwalter et al., 69–84; Keenan et al., 731–737.
14 *Several others found no change*: Christensen et al.; Crawley et al.; Casey et al.
15 *Why some tests reveal no problems*: Author's interviews with Roz Crawley, Helen Christensen, and Pamela Keenan.
15 *Back in 1956*: Leckman, "Early Parental Preoccupations," 1–26.
16 *Pregnancy does shrink your brain*: Oatridge et al., 19–26.
16 *"Baby . . . Is Eating My Brain Cells"*: Brett and Baxendale, 339–362.
17 *Some picturesque brain basics*: See, for example, Ratey, 9.
18 *Estrogen is a brain tonic*: Sherwin, 527–534, and Janowsky, 467–473.
19 *The hormone is known to play a role:* Woolley and McEwen, 5792–5801.
19 *And in neurogenesis:* Tanapat et al., 5792–5801.
19 *Researchers theorize that another hormone, progesterone:* Brett and Baxendale, 69–84.
19 *Still another camp of experts:* Brett and Baxendale, 339–362; Buckwalter, 69–84.
19 *One recent preliminary finding:* Vanston and Watston.
19 *Thanks to researchers:* Eidelman et al., 764–767.
20 *I Wish Someone Had Told Me:* Barrett.
21 *Some form of the "baby blues":* Hrdy, 170–173.
21 *Scientists in recent years:* Reynolds, 831–835.
21 *"To tamper with their equilibrium":* "The Real Victims of Sleep Deprivation." *BBC News Online Magazine* (Jan. 8, 2004); http://news.bbc.co.uk/1/hi/magazine/3376951.stm.
22 *If you persist in your slumber deficit:* Harrison and Horne, 236–249.
23 *The two studies that have made the strongest case:* Keenan et al., 731–737; Buckwalter et al., 69–84.
24 *Larger studies from Australia:* Christensen et al., 7–25; Crawley et al., 69–84; Casey, 65–76.
24 *As Helen Christensen says, "Pregnancy brain is a myth":* Cutting, 1.
25 *Researchers at Charles Sturt University:* Casey, 298–304.
25 *Finally, in another small study:* Crawley et al., 7–25.
28 *"Whatever mix of happiness and sorrow":* Ruddick, 89.
29 *A violinist's fingers:* See, for example, Sharon Begley, "Parts of The Brain That Get Most Use Literally Expand and Rewire On Demand," *Wall Street Journal Online,* Oct. 11, 2002; http://webreprints.djreprints.com/606120211414.html.
29 *In London taxi-drivers:* Maguire et al., 4398–4403.
29 *A phenomenon known as "enrichment":* Author's interviews with Fred Gage and Marian Diamond.
30 *"A chicken is only an egg's way":* Butler, in Bartlett, 755b.

Chapter 3: The Nearly Uncharted Wilderness of Mothers' Brains

31 *"She has a head":* Cited in Ehrenreich and English, 65.
31 *Inside the huge metal cylinder:* This description is based on my visit to Yale in Sept. 2003, and interviews with James Leckman, James Swain, and Tara Magnuson.

32 *Workings of the 300 million feet of wiring:* Sandra Blakeslee, "How Does the Brain Work?" *New York Times,* Nov. 11, 2003, F4.

32 *Once used purely for medical purposes:* Parry and Matthews (available at http://www.fmrib.ox.ac.uk/fmri_intro/fmri_intro.htm).

34 *"We are . . . delving into neurobiological factors":* Numan, viii.

34 *The new wave of research is actually a niche:* See, for example, Emily Eakin, "I Feel, Therefore I Am," *New York Times,* April 19, 2003, D7.

34 *The Institute for Research on Unlimited Love:* Institute's Web site: http://www.unlimitedloveinstitute.org/.

35 *A 1968 paper:* Bell, 81–95.

35 *Mothers tend to look:* Collis and Schaffer, 315–320.

35 *Babies will be the first:* Dillon, 267–275.

36 *Male scientists determinedly preferred:* Author's interview with Marian Diamond.

36 *In the field of stress physiology:* Taylor, 18.

37 *Diamond published a paper:* Diamond, "Brain Plasticity Induced by Environment and Pregnancy," 171–178.

38 *As Susan Faludi wrote:* Faludi, 327.

40 *"For more than 180 million years":* Maclean, "Women: A More Balanced Brain?" 422.

40 *The "triune," or three-in-one brain:* See, for example, Sagan, 57–59.

41 *In a series of careful experiments:* Numan, 155–176.

42 *Lorberbaum had brain-scanned:* Author's e-mail communication with Lorberbaum, Nov. 14, 2004; also see Lorberbaum 2002, 431–445.

42 *The pecan-sized brain of the rat:* I have respectfully borrowed the pecan/cantaloupe analogy from Diamond's "Male and Female Brains"; http://www.newhorizons.org/neuro/diamond_male_female.htm.

42 *In 2002, James J. Dillon:* Dillon, 267–275.

44 *Fathers appear to have a very different:* J. P. Lorberbaum et al., poster.

44 *Donald Symons, an anthropologist:* Steven Johnson, 115–116.

45 *At University College, London:* Bartels and Zeki, 1155–1166.

45 *Rat mothers seem to share:* Description of Joan Morrell's work is based on the author's personal communication with her by phone and e-mail in Feb. 2004.

47 *Women pregnant with boys tend to eat:* Study, "Women Pregnant with Boys Eat More," Associated Press, June 6, 2003.

47 *Heal faster when wounded:* Society for Neuroscience press release, November 2003.

47 *One of the most remarkable changes:* Seifritz, 1367–1375.

48 *Any woman "who relied on looks alone":* Hrdy, "Mother Nature," 24.

Chapter 4: Perception

51 *"The smell and taste of things":* Proust, cited in Bartlett, 9076.

51 *"Check this out":* This scene was based on my visit with Craig Kinsley at his lab in Oct. 2003.

53 *Clinical scientists at the University of Calgary:* This section is based on Shingo et al., 117–120, and telephone interviews in 2004 with Samuel Weiss and Robert Bridges.
54 *Sheep will neglect their newborns:* Hrdy, 157.
54 *A tiny pair of pits:* Furlow, 38–45.
54 *Biochemical bouquets:* Ibid.
54 *Tests have shown that women can distinguish:* Motluk, 36–40.
54 *Women also tend to have a particularly heightened sense of smell:* Author's interview with Charles Wysocki, Monell Chemical Senses Center, Philadelphia. See also Dalton et al., 199–200.
55 *Margie Profet:* see, for example, M. Nesse and George C. Williams, "Evolution and the Origins of Disease," http://www.direct-ms.org/articles/Evolution-OriginsOfDisease.pdf.
56 *The heightened sensitivity:* Discussion of women's enhanced sense of smell based on: "New Mothers Protected from Stress, Have Heightened Reward, Smell," Society for Neuroscience Press release, Nov. 2003; and phone and e-mail interviews with Daniel Broman of Umea University, Sweden.
56 *Imprinting is critical for ewes:* Hrdy, 157–158.
56 *Within one week after birth:* Porter et al., 151–154.
56 *Fleming isn't convinced:* Author's interview with Alison Fleming, 2004.
57 *The female brain starts out more sensitive to sounds:* Fisher, 86.
57 *One study found that nearly 60 percent of mothers:* Fleming, "Plasticity of Innate Behavior."
57 *Other research, performed about a decade before:* Ibid.
57 *This seemingly "instinctive" response:* Author's interview with Alison Fleming.
58 *The manner in which most adults:* See, for example, Dillon.
60 *She separated new mothers:* Fleming, "Plasticity of Innate Behavior."
61 *A cartoon figure with distorted features:* See, for example, Sagan, 36–37.
61 *In 1994, two neuroscientists:* Xerri et al., 1710–1721.
62 *"Our deeds determine us":* Eliot, cited in Hrdy, "Mother Nature," 31.
63 *As early as 1819:* Diamond, "Enriching Heredity," 2.
63 *An adult monkey's motor cortex could change:* Holloway, 79–85.
63 *In 1993, in one of the earliest findings:* Sharon Begley, "Survival of the Busiest: Parts of the Brain That Get Most Use Literally Expand and Rewire On Demand," *Wall Street Journal,* Oct. 11, 2002, B1.
64 *People who take up juggling:* Draganski et al., 311–312.
66 *A "zealot" of plasticity:* Holloway, 79–85.
66 *Jeffrey Schwartz, at the University of California:* See, for example, the Jeffrey Schwartz Web page: http://www.ocdcentre.com/pages/schwartz/art-neuro.htm.

Chapter 5: Efficiency

67 *"I have a brain":* Pat Schroeder, http://womenshistory.about.com/cs/quotes/a/pat_schroeder.htm.
67 *Early one summer morning:* This scene is based on telephone interviews with Kathy Mayer.

70 *The Yerkes-Dodson law:* See, for example, http://www.webster-dictionary.org/definition/Yerkes-Dodson%20law.

70 *A mental skill that the Harvard psychiatrist John Ratey:* Ratey, 114. Subsequent quotes are from email and telephone interviews in 2004.

71 *The British novelist Rachel Cusk:* Cusk, 157.

72 *The famed author of the Harry Potter series:* Margaret Weir, "Of Magic and Single Motherhood," *Salon.com,* March 31, 1999.

72 *"Teletubbies cake where?":* Pearson, 71.

73 *The neuroanatomist Marian Diamond notes:* Diamond, "Male and Female Brains."

74 *Women do seem to use more of their brains at once:* Hardin, "News Flash: Men Do Hear—But Differently Than Women, Brain Images Show," press release, Indiana University School of Medicine, November 28, 2000.

74 *Called the corpus callosum:* See, for example, Blum, 47–48; and author's interviews with Jeffrey Lorberbaum and Mark George.

74 *Helen Fisher,* a *Rutgers University anthropologist:* Fisher, 3–28.

75 *In the rat studies:* Kinsley et al., "Motherhood Improves Memory and Learning," 137–138.

75 *Since then, one of Lambert's undergraduate students:* A. Garrett et al., poster.

76 *Jay Leno joked:* The NBC press office has confirmed the *Tonight Show* transcript.

76 *Lambert . . . even designed one rat trial:* Lambert, San Diego poster, 2004; author's interview with Kelly Lambert.

77 *Catherine Woolley, a neurobiologist:* Woolley, 2549–2594.

77 *And John Morrison . . . has found:* Tang, 215–223.

77 *"If estrogen can help females":* John Morrison interview with Norman Swan, on the Australian Broadcasting Corporations "Health Report," July 8, 2002.

77 *Anne Lamott once compared:* Lamott, *Operating Instructions,* 80.

78 *The Virginia researchers found support:* Kinsley et al., "Motherhood Improves Memory and Learning," 137–138, and author's interviews with Craig Kinsley and Kelly Lambert.

79 *When James Dillon:* Dillon, 277–275.

80 *"With a baby, much of your time":* Stern, 13.

81 *In 1998, Gage made neuroscience history:* Park, http://www.time.com/time/archive/preview/0,10987,997673,00.html.

81 *Kinsley and Lambert tested:* Kinsley et al., poster, and author's interviews with Craig Kinsley and Kelly Lambert.

82 *the "grandmother hypothesis":* Fisher, 184–185, and author's interview with Kristen Hawkes.

83 *As "Maternal Thinking" author Sara Ruddick . . . writes:* Ruddick, "What Do Grandmothers Know and Want?"

Chapter 6: Resiliency

85 *"As pursed lips clamp":* Hrdy, *Mother Nature,* 536.

85 *Motherhood, . . . "made me join the human race":* Erica Jong confirmed this quote in an e-mail to me.

86 *The famous Nurses' Study:* Achat et al., 735–750.

86 *And a separate University of Michigan study:* See, for example, BBC News, "Chatting 'Good for the Brain,'" Oct. 23, 2001.

87 *Oxytocin was discovered . . . by Sir Henry Dale:* Uvnas-Moberg, *The Oxytocin Factor,* 3.

87 *Might one day be used as an antidepressant:* Author's interviews with Cort Pedersen, Karen Parker, and Paul Zak, 2004.

87 *Oxytocin acts in two ways:* Numan, 194–209.

87 *Male prairie voles given a shot of the stuff:* Egan, http://www.whsc.emory.edu/_pubs/em/1998summer/vole.html.

87 *Mice bred so that they can't be:* Ferguson, 284-288.

87 *Rats respond:* Uvnas-Moberg, *The Oxytocin Factor,* 66.

87 *In humans, oxytocin levels peak:* Ibid., 118.

87 *"the endocrinological equivalent of candlelight":* Hrdy, *Mother Nature,* 154.

88 *New receptors for the hormone:* Ibid., 138.

88 *At Sweden's Karolinska Institute:* Uvnas-Moberg, *The Oxytocin Factor,* 93–103, and Uvnas-Moberg, "Oxytocin May Still Mediate the Benefits," 819–835.

88 *Stressed ten lactating and ten non-lactating:* Altemus, 2954–2959.

89 *The neurons that produce oxytocin in the brain of a rat:* Theodosis, 9-58.

89 *Rats exposed to extra oxytocin:* Holst, 85-90.

89 *A team of researchers led by Kazuhito Tomizawa:* Tomizawa et al., 384–390.

90 *"Readers of either sex now have an additional reason":* Monks et al., 327–328.

90 *Rutgers behavioral neuroscientist Tracey Shors:* Daniel DeNoon, "Superwoman's Secret Identity": Mom, *WebMD Medical News,* Nov. 14, 2003.

90 *Shors believes this "unique response system":* "Positive Benefits of Pregnancy and Motherhood: Heightened Sensitivity to Smell and Reward; and Improved Wound Healing and Protection Against Stress," Society for Neuroscience press release, Nov. 11, 2003.

90 *Glucocorticoids are one reason that sustained stress:* Sapolsky, *Why Zebras Don't Get Ulcers,* 240–247.

91 *A team of German scientists:* Heinrichs et al., 1389–1398.

91 *Ever since the early 1930s:* Sapolsky, *Why Zebras Don't Get Ulcers,* 1–19.

93 *This reaction calls more on the parasympathetic system:* Uvnas-Moberg, *The Oxytocin Factor,* x.

93 *In a widely noted paper published in the year 2000:* Taylor et al., "Behavioral Responses to Stress in Females," 411–429.

94 *According to research by Rena Repetti:* Unpublished paper presented at the biennial meeting of the Society for Research in Child Development, April 1997, and author's interview with Repetti.

95 *Daniel Stern . . . calls the "motherhood mindset":* Stern, 3.

96 *In a study Stern conducted in Boston:* Ibid., 134.

96 *"weakness admitted is the very stuff of good friendships":* Weldon, 34.

97 *In a sixteen-year study of wild baboons:* Silk et al., 1231–1234.

98 *"One of the most reliable sex differences there are":* Taylor, *The Tending Instinct,* 24.

98 *In this study, Larry Jamner:* Unpublished ms. and e-mail correspondence with Larry Jamner.

99 *But Barry Keverne:* Author's telephone interview, October 2004, and see Keveren's Web site: http://www.zoo.cam.ac.uk/zoostaff/keverne.htm.

99 *And a remarkable experiment in 2003:* Pilcher, http://www.nature.com/news/2003/031110/full/031110-7.html.

100 *"Across the board, higher-trust countries":* See also Paul Zak's Web page: http://fac.cgu.edu/~zakp/pubs.htm.

101 *Many studies have established that social ties:* See, for example, S. Cohen, 1940 to 1944.

Chapter 7: Motivation

103 *"Because of deep love":* Bartlett, 75a.

103 *Olivia Morales stands out:* Olivia Morales is a pseudonym for a woman I have come to know and interviewed several times. All other details of her story are factual.

104 *During the past three decades, thousands:* Author's interview with Pierette Hondagneu-Sotelo, 2003.

104 *Motivation is a key component of "emotional intelligence":* Goleman, 43.

104 *The "master aptitude":* Ibid., 78–79

104 *In a 1930 experiment:* Nissen, 377–393.

105 *Tanja MacKenzie:* Author's e-mail interview with Tanja MacKenzie, 2004.

106 *State Farm Insurance:* The information on automobile insurance policies was verified with a lawyer for the California Department of Insurance.

107 *Refers to this practice as "task ownership":* Stern, 15–17.

108 *When Susan Calleymore's twenty-six-year-old son:* Joe Garofoli, "Anti-war Alameda Woman's Trip to See Son Serving in Iraq Has Surprises for Both," *San Francisco Chronicle*, April 3, 2004.

109 *To test this hypothesis:* Wartella et al., 373–381.

110 *Prolactin was first identified in the early 1930s:* Hrdy, *Mother Nature,* 133.

110 *Prolactin works wonders:* Torner et al., 3207–3214.

110 *A female rat that has not been pregnant:* Author's interview with Alison Fleming.

111 *Male rodents and primates:* See for example, Hrdy, 86–89.

112 *The reality, maintains Sara Ruddick:* Ruddick, 1989, 160.

112 *One small but provocative study:* Mastrogiacomo, 1982–1983.

114 *"We'd grown tired of knocking on doors":* Web page: Asociacion Madres de Plaza del Mayo, http://www.madres.org/.

115 *A mother's ambition was "an integral part":* Hrdy, 110.

115 *The 2000 U.S. Census found:* U.S. Census Bureau of Household and Family Statistics, 2000.

115 *When the* New York Times *published:* Faludi, 82.

116 *Yet, as the writer Susan Faludi:* Ibid., 87.

116 A New York Times Magazine *cover story in 2003:* Belkin.

116 *A* Time *magazine cover titled:* "The Case for Staying Home."

116 *Among African American mothers:* Rosenwein, 2001.

116 *"Mom Overboard!":* Singer, 65.

117 *"I was rolling my eyes":* I interviewed Rhonda Staudt by telephone in the spring of 2004 after reading Peter Senge's account of a meeting with her. See Senge, 29.

119 *Napoleon Bonaparte once wrote:* Thoreau, 338.

Chapter 8: Emotional Intelligence

121 *"Imagining motherhood opens the door":* Brown, 15.

121 *In the mid–1990s:* Hall, 46.

122 *The idea that we have "multiple intelligences":* Goleman, 37–43.

122 *Seven years later, Peter Salovey:* Ibid., 42–46.

122 *"Love as well as other positive emotions":* Richard Davidson's quotes are from a transcript of a Canadian TV network interview (CBC) recorded in 2003.

123 *In one group of indigenous Bolivians:* Scheffler and Lounsbury, 44.

124 *In a remarkable experiment involving a miniature fMRI device:* "New Mothers Protected from Stress, Have Heightened Reward, Smell," Society for Neuroscience press release, Nov. 2003; and author's interview with Craig Ferris, Dec. 2003.

126 *At two weeks after delivery, mothers on average report:* Leckman et al., 1–26.

127 *One study, involving six-month-old babies in Japan:* Masataka, 241–246.

127 *In contrast, a test of depressed mothers:* Bettes, 1089–1096.

127 *Anne Lamott has described:* Anne Lamott, "Mother Rage: Theory and Practice," *Salon.com,* Oct. 29, 1998.

128 *When researchers used a harness:* Preston, 1–20.

128 *In a separate experiment:* Ibid.

129 *If a person sets his face:* See, for example, Goleman, 115.

129 *Using brain scans, a team:* "UCLA Imaging Study Reveals How Active Empathy Charges Emotions: Physical Mimicry of Others Jump-Starts Key Brain Activity," *Science Daily* (UCLA Press Release), April 8, 2003.

130 *Tania Singer, a researcher at University College*: Daniel Kane, *Science,* http://www.msnbc.msn.com/id/4313263/.

131 *In the early 1970s:* Lewis and Rosenblum, 67.

132 *Another Harvard study in the 1970s:* Rosenthal, 308–312.

134 *Kids' greater impulsiveness:* Wallis, May 2004.

135 *Mothers of two- to three-year-olds:* Ambert, 69.

135 *An unlucky young railroad worker:* See, for example, Ratey, 231–232.

135 *Buddhists, for instance:* Author's interview with Paul Ekman.

136 *Edward Tronick, a developmental psychologist:* Tronick, 2003.

136 *Joan Didion long ago wrote:* Didion, 11.

137 *Martin Seligman:* Goleman, 89.

138 *Paul MacLean, the NIH researcher, has argued:* MacLean, 421–439.

138 *The evolutionary leap that made:* Hrdy, *Mother Nature,* 140.

139 *Barry Keverne, at Cambridge University:* Ibid., 143–144.

Chapter 9: Mr. Moms and Other Altruists

143 *"[A] good part of what gives motherhood":* Churchwell, 285.

143 *Minutes after their babies were born:* Kerstin Uvnas-Moberg described this unpublished study in a telephone interview in early 2004.

144 *Altruism, defined in* Webster's: *Merriam-Webster's Collegiate Dictionary,* 34.

144 *In* Aging Well: Stephen G. Post, "Studies on Love and Kindness Get Better with Age," *Science & Theology News*, February 2004. http://www.stnews.org/archives/2004_february/altr_studies_0204.html.

144 *And a 2003 report:* Carey Goldberg, "For Good Health, It Is Better to Give, Science Suggests," *Boston Globe,* Nov. 28, 2003.

144 *That the psychologist Erik Erikson characterized:* See, for example, McAdams and de St. Aubin, 62, 103–115.

146 *Compared to roughly 90 percent of the mammal world:* Numan and Insel, 1.

147 *A mail clerk whom Coltraine interviewed:* Coltrane, *Family Man,* 119.

149 *Groundbreaking research by two Canadian:* Abrams; see http://cms.psychologytoday.com/articles/pto-20020301-000025.html.

151 *The kinder, gentler hormone levels:* Ibid.

151 *Comparing human fathers and nonfathers:* Fleming, "Testosterone and Prolactin," 399–413.

151 *As the father cares:* Author's interview with Professor Robert Bridges of Tufts University, 2004.

151 *Studies of prairie voles:* Wang, 111–120. Alan Zarembo, "Swingers May Be Slaves to Genes," *Los Angeles Times*, Thursday, June 17, 2004.

152 *A human father's circulating hormones:* Abrams.

152 *125,000 children were adopted:* See, for example, the Web site of the nonprofit research-oriented Evan B. Donaldson Adoption Institute, http://www.adoptioninstitute.org/FactOverview/domestic.html.

153 *Marmoset siblings that carry:* Roberts, 713–720.

154 *In 1985, researchers:* Numan and Insel, 318.

154 *Michael Numan at Boston College:* Ibid., 319.

155 *Tens of thousands of children:* http://www.adoptioninstitute.org/FactOverview/domestic.html.

155 *"You taught me how to care":* "The Things They Wrote," *New York Times,* Nov. 11, 2003, A23.

156 *In San Francisco, Suzan Houseman:* Jose Antonio Vargas, "New Life Began at Age 44," *San Francisco Chronicle,* Nov. 27, 2003, 1A.

157 *At a center for homeless people:* Elizabeth Share described this scene to me after visiting the Shanti Project homeless center in San Francisco. I later confirmed the details with Shanti employees and interviewed Alyssa Nickell by telephone.

158 *The biologist William Hamilton:* See, for example, "A Tribute to W. D. Hamilton," *Times* (London), March 9, 2000, and David Jay Brown's interview with Robert Trivers and David Jay Brown's interview with Robert Trivers at http://www.mavericksofthemind.com/tri-int.htm.

158 *Researchers who have scanned:* J. Rilling et al., 395–405.

159 *Elderly patients in a nursing home:* See, for example, Sapolsky, *Why Zebras Don't Get Ulcers,* 269.

160 *"I'm really happy":* Jennifer Byrne, "Guatemala, Out of the Dump," ABC-TV broadcast, June 15, 2003.

Chapter 10: Better Than Business School

161 *"If you can manage a group":* Ekrut, 79.

161 *There are nearly 26 million working mothers:* U.S. Bureau of Labor Statistics news release, "Employment Status of the Population by Sex, Marital Status and Age of Own Children Under 18," http://stats.bls.gov/news.release/famee.t05.htm.

162 *The "face-to-face economy":* News Hour with Jim Lehrer, August 16, 2004.

162 *Joanne Hayes-White, the first:* Ryan Kim, "Newsom Names Female Fire Chief: Historic Choice for Big-City Department," *San Francisco Chronicle,* Jan. 11, 2004.

163 *Earn about ninety cents to the dollar:* Caryl Rivers and Rosalind Chait Barnett, "Wage Gap for Working Mothers May Cost Billions," *Women's E-News,* July 3, 2000.

163 *Princeton University researchers:* Cuddy and Fiske.

163 *Fatherhood usually* helps *their careers:* Mason and Goulden.

164 *U.S. Air Force recruiters:* Cary Cherniss, "The Business Case for Emotional Intelligence," Consortium for Research on Emotional Intelligence in Organizations, http://www.eiconsortium.org/research/business_case_for_ei.htm.

165 *Madeleine Albright:* Rowe-Finkbeiner; http://www.findarticles.com/p/articles/mi_m0838/is_2003_March-April/ai_100807068/pg_2.

165 *A Wellesley College in-depth report:* Ekrut.

166 *Ann Moore, CEO:* Interview cited in Crittenden, 26.

167 *One of the most audacious estimates:* Annelena Lobb, "Does Mom Need a Raise?" *CNNMoney,* May 8, 2002.

169 *Who work shifts:* "Married Workers Fare Better at Shift Work than Singles," press release from the Society for Industrial and Organizational Psychology, March 30, 2004.

169 *Arlie Hochschild, a sociology professor:* Hochschild.

170 *In Ann Crittenden's book:* Crittenden, 151.

170 *Secretary of State Albright:* Rowe-Finkbeiner.

170 *"Offices do for some":* Weldon, 110.

172 *Crittenden tells:* Crittenden, 7–8.

172 *Mary Parker Follett:* Ekrut, 1.

173 *Sally Helgesen's* The Female Advantage*:* Helgesen.

174 *After six kids:* Dahle, 60.

175 *"We may not notice the amount":* Bateson, 141.

176 *Judith Matloff:* Matloff.

Chapter 11: Smarter Than Ever

179 *"Sometimes I feel that my mother":* Cook, 98.

179 *Marian Sandmaier knew:* Marian Sandmaier, "Listening to Zebras," *Washington Post,* June 3, 2003.

181 *Advertised graduates such as one four-year-old:* See description of Michelle Gauger at Web site of The Institutes for the Achievement of Human Potential, http://www.iahp.org/well/children/gaugers.html.

181 *In 1997, nervous parents passed around:* Nash.

182 *A 2004 Wall Street Journal article:* Jeffrey Zaslow, "Is Belly Talk Helpful or Just Plain Crazy?" *Wall Street Journal,* April 29, 2004.

183 *In the late 1990s, U.S. mothers:* Author's interview with Sharon Hays; also see Kathleen Gerson, "Work Without Worry," *New York Times,* May 11, 2003.

183 *A big reduction in "leisure time":* Andrew J. Cherlin and Prem Krishnamurthy, "What Works for Mom," *New York Times,* May 9, 2004.

184 *The switch has come about:* Stearns, 14–24.

185 *Perchlorate, a toxic component:* Glen Martin, "Rocket Fuel Found In Milk In California: Not Clear If Amount Imperils Children," *San Francisco Chronicle,* June 22, 2004.

185 *The number of overweight kids:* See, for example, prepared remarks of U.S. Surgeon General Vice Admiral Richard H. Carmona, to the California Childhood Obesity Conference, Jan. 6, 2003, http://www.surgeongeneral.gov/news/speeches/califobesity.htm.

185 *The World Health Organization has attributed:* International Obesity Task Force; http://www.iotf.org.

186 *The 2004 scandal over doctors' being paid:* See, for example, Gardiner Harris, "Pfizer to Pay $430 Million Over Promoting Drug to Doctors," *New York Times,* May 14, 2004.

186 *More than 40 million Americans:* Marian Sandmaier, "Listening to Zebras," *Washington Post,* June 3, 2003.

187 *Ritalin production soared:* Marilyn Gardner, "Be All That You Can Be—or More: The Drug Industry Has Something Just for You," *Christian Science Monitor,* April 17, 2003.

187 *In 2000, close to 60 percent of American parents:* Stearns, 91.

187 *Some cities, such as Alexandria:* Ibid., 93.

188 *In the San Francisco County School District:* These statistics are courtesy of San Francisco Unified School District press office.

189 *Spending on average just short of twenty hours:* Nielsen Media Research, 2000 statistics, cited on Web site of National Center for Children Exposed to Violence, http://www.nccev.org/violence/statistics/statistics-media.html.

189 *70 percent of daycare centers use television:* Billy Tashman, "Sorry Ernie, TV Isn't Teaching," *New York Times,* Nov. 12, 1994.

190 *Almost two out of three television programs:* The National Television Violence Study found that nearly two out of three television programs contained some violence, averaging about six violent acts per hour: See "Key Facts: TV Violence," Kaiser Family Foundation (Spring 2003), http://www.kff.org/entmedia/3335-index.cfm.

190 *Spending for advertising and marketing:* Jolayne Houltz, "Companies Sell Details on Millions of Children," *Seattle Times,* July 6, 2004.

190 *The so-called consumer culture:* See the Web site of Center for a New American Dream, http://www.newdream.org/kids/borntobuy.php.

191 *Bowlby stressed what he described:* Hrdy, *Mother Nature,* xiii.

191 *"If they knew that it was O.K.":* Gladwell.

192 *A panel of academically distinguished:* A description of the report can be found at http://www.americanvalues.org/html/hardwired.html.

192 *And the Motherhood Project:* See http://www.watchoutforchildren.org/html/about_us.html.

193 *Mark Flinn, a professor of anthropology:* Small, 2000.

194 *The* New York Times *published a chart:* Cherlin and Krishnamurthy, "What Works for Mom."

Chapter 12: Reengineering the Mommy Track

197 *"Mothers' passionate concern":* Smith, 7.

198 *Workers lucky enough to have jobs: National Study of the Changing Workforce,* The Families and Work Institute, 2003.

198 *"There's trouble in the engine":* Iris DeMent, "Quality Time," from the album, *That's the Way I Should,* Warner Brothers, 1996.

199 *In 2002 . . . 72 percent of mothers:* U.S. Census Bureau, "Fertility of American Women: June 2002," October 2003.

199 *Families in which both members of married couples:* Tamar Lewin, "Now a Majority: Families with 2 Parents Who Work," *New York Times,* Oct. 24, 2000.

199 *Women made up half the student bodies:* Mason and Goulden; http://www.grad.berkeley.edu/deans/mason/index.html.

199 *Its annual list of the Top 100 Companies:* An interactive list of the Top 100 firms for working mothers can be found at http://www.workingwoman.com/bestlist.html.

201 *Jennifer Griffin, a Jerusalem-based television reporter:* Matloff.

201 *The median hourly wage of caregivers:* I drew this comparison from the 2001 National Occupational Employment and Wage Estimates provided by the U.S. Department of Labor Bureau of Labor Statistics, http://www.bls.gov.

201 *What makes being a mammal so painful:* Lambert, "The Life and Career of Paul Maclean."

202 *A full 49 percent of employers:* The Child Care Partnership Project Employer Toolkit, U.S. Department of Health and Human Services, http://nccic.org/ccpartnerships.

202 *When they're down with the flu:* G. W. Johnson, "Sick Child Care: An Idea Whose Time Has Come," presentation at the Emergency Child Care Conference, Indianapolis, 1997.

202 *The giant computer firm:* http://www.workingwoman.com/bestlist.html.

202 *As Ravenna Helson . . . has noted:* Paris and Helson, 172–185, and author's interview with Helson.

203 *"aching desire to be with their children":* de Marneffe was quoted in the *Publisher's Weekly* review of *Maternal Desire,* http://www.amazon.com/exec/obidos/ASIN/0316059951/bridgebooks/103–4940296–1704660.

204 *About 24 percent of unscheduled absences:* CCH Incorporated Unscheduled Absence Survey 2004; see http://www.cch.com/default.asp.

204 *In one survey, 56 percent:* Flexible Resources, Client Profile: Organizations Employing Professionals in Flexible Work Arrangements, 1999.

204 *One study of 11,815 managers:* Rivers and Barnett, "Wage Gap for Working Mothers May Cost Billions."

204 *Less than 25 percent of tenure track faculty:* Author's interview with Mary Ann Mason.

205 *Working an average of ninety-four hours per week:* Mason and Goulden.

205 *A year-long "stop the clock" procedure:* Author's interview with Mary Ann Mason.

205 *Harvard University may have erred:* http://www.workingwoman.com/bestlist.html.

205 *The medical school offers fifty fellowships:* The program is called the 50th Anniversary Program for Scholars in Medicine, and was established in 1995 to celebrate the 50th anniversary of the admission of women to the Medical School. More information can be found at http://www.news.harvard.edu/gazette/2002/10.03/20-medsholars.html.

206 *The accounting firm Deloitte:* Belkin.

206 *Marriott International Inc.:* http://www.workingwoman.com/bestlist.html.

206 *Taking parental leave at Ernst & Young:* Belkin.

207 *The Balancing Act:* Joan Ryan, "Woolsey Puts Focus on Family," *San Francisco Chronicle,* Jan. 18, 2004.

208 *In a May 2004 address:* See Woolsey press release at http://www.woolsey.house.gov/newsarticle.asp?RecordID=295.

208 *Only two rich nations:* Stadtman, "The Least Worse Choice."

209 *George W. Bush revoked:* Ibid.

209 *California became the first state:* "New State Laws," *San Francisco Chronicle,* December 31, 2002.

209 *Found that 25 percent of employees:* "Employers Must Shoulder Some Blame for Shortage of Qualified Workers," Radcliffe Advanced Institute for Advanced Study report cited in *Ideas and Trends,* CCH, Inc., May 3, 2000.

209 *One family-owned hosiery company:* Ibid.

209 *Three in five preschoolers:* See Web site of Children's Defense Fund, http://www.childrensdefense.org/.

210 *Two in five disadvantaged preschoolers:* Ibid.

210 *You'd have to wait two years:* Prince William County officials confirmed this in a telephone interview in September 2004.

210 *Nixon dealt this movement:* Douglas and Michaels, 245.

210 *The power of breastfeeding:* See, for example, several studies cited in the American Academy of Pediatrics Work Group on Breastfeeding, 1035–1039.

210 *The American Academy of Pediatrics has recommended:* Ibid.

211 *Just 12.5 percent of full-time working moms:* Ann Scott Tyson, "Congress Looks at Breast-Feeding and the Workplace," *Christian Science Monitor*, March 31, 1998.

212 *William James:* See, for example, http://www.emory.edu/EDUCATION/mfp/said.html.

212 *More than 5,000 mothers:* Casey et al., 298–304.

212 *A strong relationship between poverty and child-abuse:* See, for example, Drake and Pandey, 1003–1018.

212 *Kristen Brunson:* Brunson, "Life-Long Progressive Impairment," poster.

212 *Rick Hanson . . . has proposed:* Hanson et al., 17–18.

Chapter 13: Political Drive

215 *"We are used to seeing what we call 'a mother':* Gilman, 69.

216 *Paul MacLean described:* MacLean, "Women: A More Balanced Brain?" 421–439.

217 *Senator Hillary Clinton noted:* Hillary Rodham Clinton, "Now Can We Talk About Health Care?" *New York Times,* April 18, 2004.

217 *The novelist Barbara Kingsolver complained:* Barbara Kingsolver, "No Glory in Unjust War on the Weak," October 14, 2001, http://www.zmag.org/kingsolver.htm.

217 *U.S. temperance activists:* Pictures can be viewed in an online exhibit of the National Women's History Museum, http://www.nmwh.org/exhibits/gallery_12.html.

218 *Even suffragettes printed:* Ibid.

218 *The feminist leader Jane Addams:* see Halsall, "The Internet Modern History Sourcebook," http://www.fordham.edu/halsall/mod/modsbook.html#Sources%20of%20Material%20Here.

218 *In the name of "social-housekeeping":* Author's e-mail interview with Theda Skocpol, Oct. 8, 2003.

219 *"Many thought of themselves as mothers":* Ibid.

219 *"A woman is a nobody":* Ward and Burns, 42.

219 *"miserable little underdeveloped vandals":* Ibid., 51.

219 *As a newlywed in 1840:* Ibid., 30.

220 *The U.S. floral and greeting card industries:* Ruth Rosen, "Mothers Doing What?" *San Francisco Chronicle,* May 8, 2003.

220 *The Greeks held festivals:* This and much of the following description of the origins of Mother's Day comes from Cathryn Meurer, "Mom-orabilia: Mother's Day History from Rea to the Soccer Mom," *CNN Interactive*, May 3, 1999.

221 *When Candace Lightner:* Lord.

221 *Recurring health problems spurred Lois Gibbs:* Lois Gibbs won the annual Goldman Environmental Prize in 1990. These details are from the Goldman Foundation's profile of her, and can be found at http://www.goldmanprize.org/recipients/recipientProfile.cfm?recipientID=15.

222 *Mothers Acting Up:* http://www.mothersactingup.org.

222 *As if echoing Addams, C. C. Pelmas:* Author's e-mail correspondence from C. C. Pelmas, Sept. 12, 2003.

222 *Ruddick reaffirmed her hope:* Ruddick, "Making Connections Between Parenting and Peace."
222 *Ancient city-state of Sparta:* Elshain, xiii.
222 *Mothers, alas:* Jetta, 230.
223 *As Virgina Woolf has written:* Ibid.
223 *Voted for the Democratic Party challenger:* E-mail correspondence from Alysia Snell, a partner in Lake, Snell, Perry & Associates, Inc., polling firm.
223 *Bush's strategist, Karl Rove:* Tumulty and Novak.
223 *"Since 9/11":* Ibid.
223 *(They are also more reliable voters):* July 8, 2003, memo by Celinda Lake, of Lake, Snell, Perry & Associates, Inc.
223 *This rang true:* Ellen Goodman, "The Myth of 'Security Moms,'" *Boston Globe,* October 7, 2004.
224 *Born in Cork:* Gorn, 7.
225 *"It freed her":* Ibid., 5.
225 *The Million Mom March:* http://www.millionmommarch.org/.
225 *When congressional Democrats:* Schroeder, 75.
225 *Governor Jennifer Granholm of Michigan: Detroit Free Press* editors, "Q & A with Gov. Jennifer Granholm: State Budget Forces Tough Choices and Priorities," *Detroit Free Press,* February 20, 2003.
226 *Patty Murray's image:* Sheryl Gay Stolberg, "Working Mothers Swaying Senate Debate, as Senators," *New York Times,* June 7, 2003.
227 *Enola Aird, a Panamanian-born:* William Rasberry, "Mother Load," *Washington Post,* Nov. 18, 2002.
228 *The chairman of CBS:* This is posted at http://www.watchoutforchildren.org/html/letter_to_cbs.html.
228 *Fueling this concern:* Judith Stadtman Tucker, "An Interview with Enola Aird," *The Mothers Movement Online,* June 2003.
229 *Brundage was:* Author's interview with Joanne Brundage; details at Mothers and More Web site, http://www.mothersandmore.org/AboutUs/history.shtml.
229 *Mothers Ought to Have Equal Rights:* See http://www.mothersoughttohave equalrights.org/.
230 *Activists in more than twenty states:* Chapko and English; http://www.find articles.com/p/articles/mi_m3495/is_12_47/ai_95679831.
231 *Charlotte Perkins Gilman offered:* Gilman.

Chapter 14: Neuroscientists Know Best

235 *Alison Fleming notes:* Fleming, "Plasticity of Innate Behavior."
235 *Advice that might be just as useful:* Hallowell and Ratey, 249.
236 *"Never wake a sleeping adult":* Ehrenreich, 148.
238 *A twenty-five-year controlled study:* See Web site of the nurse-family partnership, http://www.nccfc.org/faq.cfm.
238 *Sarah Hrdy says it's likely:* Hrdy, "Mothers and Others"; http://www.natural historymag.com/0501/0501_feature.html.

238 *Physical exercise increases circulation:* Ratey, 359.

240 *A balanced diet is demonstrably good:* Ibid., 368–371, and "From Green and Leafy to a Sharper Brain," *New York Times,* July 20, 2004.

240 *And fish really does:* See, for instance, Gorman; http://www.time.com/time/archive/preview/0,10987,998346,00.html.

240 *David Meyer, a University of Michigan cognition expert:* Martin.

241 *Cite research indicating children:* See "When Dads Clean House, It Pays Off Big Time," UC Riverside Press release, June 9, 2003, at http://www.newsroom.ucr.edu/cgi-bin/display.cgi?id=611.

241 *Moms are likely to find their husbands sexy:* Ibid.

BIBLIOGRAPHY

Abrams, Douglas Carlton. "Father Nature: The Making of a Modern Dad; It Takes a Lot More Than Testosterone to Make a Father Out of a Man." *Psychology Today* (March–April, 2002).

Achat, H., et al. "Social Networks, Stress and Health-Related Quality of Life." *Quality of Life Research,* vol. 7, no. 8 (Dec. 1998).

Allport, Susan. "A Natural History of Parenting: A Naturalist Looks at Parenting in the Animal World and Ours." *iUniverse,* 2003.

Altemus, M. "Suppression of Hypothalamic-Pituitary-Adrenal Axis Responses to Stress in Lactating Women." *Journal of Clinical Endocrinology & Metabolism,* vol. 80, no. 10 (Oct. 1995).

Ambert, Anne-Marie. *The Effect of Children on Parents.* Haworth Press, 2001.

"American Academy of Pediatrics: Breastfeeding and the Use of Human Milk." *Pediatrics,* vol. 100, no. 6 (December 1997).

Barrett, Nina. *I Wish Someone Had Told Me: A Realistic Guide to Early Motherhood.* Academy Chicago Publishers, 1997.

Bartels, A., and S. Zeki. "The Neural Correlates of Maternal and Romantic Love." *Neuroimage* (2004) (article in press).

Bartlett, John. *Familiar Quotations.* Little, Brown, 1968.

Bateson, Mary Catherine. *Composing a Life: Life As a Work in Progress—The Improvisations of Five Extraordinary Women.* Plume, 1976.

Belkin, Lisa. "The Opt-Out Revolution." *New York Times Magazine* (Oct. 26, 2003).

Bell, R. Q. "A Reinterpretation of the Direction of Effects in Studies of Socialization." *Psychological Review,* vol. 75, no. 2 (1968).

Bettes, B. A. "Maternal Depression and Motherese: Temporal and Intonational Features." *Child Development,* vol. 59, no. 4 (Aug. 1988).

Blum, Deborah. *Love At Goon Park: Harry Harlow and the Science of Affection.* Perseus Publishing, 2002.

______. *Sex on the Brain: The Biological Differences Between Men & Women.* Penguin Books, 1997.

Born, J., et al. "Timing the End of Nocturnal Sleep." *Nature,* vol. 397 (1999).

Brett, M., and S. Baxendale. "Motherhood and Memory: A Review." *Psychoneuroendocrinology,* vol. 26, no. 4 (May 26, 2001).

Brown, Kurt. *The True Subject: Writers on Life and Craft.* Graywolf Press, 1993.

Brunson, Kristen L., et al. "Life-Long Progressive Impairment in Memory and Hippocampal Synaptic Physiology After Early-Life Stress." Poster at Society for Neuroscience conference, San Diego, 2004.

Buckwalter, J. Galen, et al. "Pregnancy, the Postpartum, and Steroid Hormones: Effects on Cognition and Mood." *Psychoneuroendocrinology,* vol. 24 (1999).

Carson, Rachel. *Silent Spring.* Mariner Books, 2002.

Carter, C. S., "Neuroendocrine Perspectives on Social Attachment and Love." *Psychoneuroendocrinology,* vol. 8 (Nov. 1998).

Casey, Patrick, et al. "Maternal Depression, Changing Public Assistance, Food Security, and Child Health Status." *Pediatrics,* vol. 113 (Feb. 2004).

Casey, Paul. "A Longitudinal Study of Cognitive Performance During Pregnancy and New Motherhood." *Archives of Women's Mental Health,* vol. 3 (2000).

Chapko, Terry, and John English. "Paid Family Leave—It Could Happen to You: California Is the First—But Perhaps Not the Only—State to Require Paid Family Leave." *HR Magazine* (Dec. 2002).

Christensen, H., et al. "Pregnancy May Confer a Selective Cognitive Advantage." *Journal of Reproductive and Infant Psychology,* vol. 17, no. 1 (1999).

Churchwell, Gordon. *Expecting: One Man's Uncensored Memoir of Pregnancy.* Harper Collins, 2001.

Cohen, S., et al. "Social Ties and Susceptibility to the Common Cold." *Journal of the American Medical Association,* vol. 277, no. 24 (1997).

Collis, G. M., and H. R. Schaffer. "Synchronization of Visual Attention in Mother-Infant Pairs." *Journal of Child Psychology and Psychiatry,* vol. 16 (1975).

Coltrane, Scott. *Family Man: Fatherhood, Housework, and Gender Equity.* Oxford University Press, 1996.

______."Fathering: Paradoxes, Contradictions, and Dilemmas." Forthcoming in *Handbook of Contemporary Families: Considering the Past, Contemplating the Future,* edited by Marilyn Coleman and Lawrence Ganong. Thousand Oaks, 2004.

______."Research on Household Labor: Modeling and Measuring the Social Embeddedness of Routine Family Work." *Journal of Marriage and the Family,* vol. 62, (2000).

Conniff, Richard. "Reading Faces." *Smithsonian* (Jan. 2004).

Cook, Mariana. *Mothers and Sons: In Their Own Words.* Chronicle Books, 1996.

Crawley, R. A., et al. "Cognition in Pregnancy and the First Year Post-Partum." *Journal of Psychology and Psychotherapy,* vol. 76, (March 2003).

Crittenden, Ann. *If You've Raised Kids, You Can Manage Anything: Leadership Begins At Home.* Gotham Books, 2004.

Cuddy, Amy, J. C., and Susan T. Fiske. "When Professionals Become Mothers, Warmth Doesn't Cut the Ice." Unpublished ms., Princeton University, 2004.

Cusk, Rachel. *A Life's Work: On Becoming a Mother.* Picador, 2002.

Cutting, Tania. "Memory Loss During Pregnancy." *Australia National University Reporter,* vol. 31, no. 6 (May 19, 2000).

Dahle, Cheryl. "You, Only Better." *Working Mother* (Aug. 2003).

Dalton, P., et al. "Gender-Specific Induction of Enhanced Sensitivity to Odors." *Nature Neuroscience,* vol. 5, no. 3 (March 2002).

Damasio, H., et al. "The Return of Phineas Gage: Clues About the Brain from the Skull of a Famous Patient." *Science,* vol. 264, no. 5162 (1994).

Damp, Dennis. "Health Care Job Explosion: High Growth Health Care Careers and Job Locator." Brookhaven Press, 2001.

Dess, Nancy K. "The Nature of Monkeys and Mothering, *Psychology Today* (May–June 2001).

DeVries, A. C., et al. "Social Modulation of Stress Responses." *Physiolology & Behavior,* vol. 79, no. 3 (Aug. 2003).

Diamond, M. C. *Enriching Heredity: The Impact of the Environment on the Anatomy of the Brain.* Simon & Schuster, 1988.

______. "Male and Female Brains." Summary of annual lecture for Women's Forum West, San Francisco, 2003.

______. "Brain Plasticity Induced by Environment and Pregnancy." *International Journal of Neuroscience,* vol. 2, no. 4 (Nov. 1971).

Didion, Joan. *The White Album.* The Noonday Press, 1979.

Dillon, James J. "The Role of the Child in Adult Development." *Journal of Adult Development,* vol. 9, no. 4 (Oct. 2002).

Douglas, Susan, and Meredith Michaels. *The Mommy Myth: The Idealization of Motherhood and How It Has Undermined Women.* Free Press, 2004.

Draganski, B., et al. "Changes in Gray Matter Induced by Training." *Nature,* vol. 427 (Jan. 22, 2004).

Drake, B., and S. Pandey. "Understanding the Relationship Between Neighborhood Poverty and Specific Types of Child Maltreatment." *Child Abuse and Neglect,* vol. 11 (Nov. 20, 1996).

Egan, Kate. "Love & Sex: The Vole Story." *Emory Medicine* (Summer 1998).

Ehrenreich, Barbara. *The Worst Years of Our Lives: Irreverent Notes from a Decade of Greed.* Pantheon, 1990.

Ehrenreich, Barbara, and Deirdre English. *For Her Own Good: 150 Years of the Experts' Advice to Women.* Anchor Books Editions, 1979.

Eidelman, A. I., et al. "Cognitive Deficits in Women After Childbirth." *Journal of Obstetrics and Gynecology,* vol. 81, no. 5 (May 1993).

Eisenberger, Naomi and Matthew Leiberman. "Why Rejection Hurts: A Common Neural Alarm System for Physical and Social Pain." *Trends in Cognitive Sciences,* vol. 8, no. 7 (July 2004).

Ekrut, Sumru. *Inside Women's Power: Learning from Leaders.* Winds of Change Foundation and Wellesley Centers for Women, 2001.

Elshtain, Jean. *Women and War.* University of Chicago Press, 1987.

Eliot, George. *Adam Bede.* Penguin Books, 1980.

Ellison, Katherine. "Rats, Marmosets and You." *Working Mother* (Feb. 2003).

Faludi, Susan. *Backlash: The Undeclared War Against American Women."* Crown Publishers, 1991.

Ferguson, J. N., et al. "Social Amnesia in Mice Lacking the Oxytocin Gene." *Nature Genetics,* vol. 25, no. 3 (July 2000).

Fisher, Helen. *The First Sex: The Natural Talents of Women and How They Are Changing the World.* Random House, 1999.

Fleming, A. S. "Plasticity of Innate Behavior: Experiences Throughout Life Affect Maternal Behavior and its Neurobiology." Unpublished ms. prepared for the Dahlem Workshop on Attachment and Bonding: A New Synthesis, Berlin, 2003.

Fleming, A. S., et al. "Testosterone and Prolactin Are Associated with Emotional Responses to Infant Cries in New Fathers." *Hormonal Behavior*, vol. 42, no. 4. (Dec. 2002).

Friedan, Betty. *The Feminine Mystique.*, W. W. Norton & Co., 2001.

Furlow, F. B. "The Smell of Love." *Psychology Today* (March–April 1996).

Garrett, A., et al. "Parenting Experience Enhances Spatial Learning in Common Marmosets (Callithrix jacchus)." Poster presented at the Mother and Infant Conference held in Montreal, Canada, April, 2002.

Gilman, Charlotte Perkins. *Herland: A Lost Feminist Utopian Novel.* Pantheon Books, 1979.

Gladwell, Malcolm. "Do Parents Matter?" *New Yorker* (Aug. 17, 1998).

Glasper, E., et al. "Reproductive Experience Alters Anxiety and Learning Ability in *Peromyscus Californicus* Males and Females." Poster presented at the international Behavioral Neuroscience Society annual meeting, Cancun, Mexico, 2001.

Goleman, Daniel. *Emotional Intelligence: Why It Can Matter More Than IQ.* Bantam Books, 1995.

Gorn, Elliott J. *Mother Jones: The Most Dangerous Woman in America.* Hill and Wang, 2001.

Gorman, Christine. "How to Eat Smarter." *Time* (Oct. 20, 2003).

Hall, Stephen S. "Is Buddhism Good for Your Health?" *New York Times Magazine* (Sept. 4, 2003).

Hallowell, Edward, and John Ratey. *Driven to Distraction: Recognizing and Coping with Attention Deficit Disorder from Childhood through Adulthood.* Touchstone, 1994.

Hanson, Rick, et al. *Mother Nurture: A Mother's Guide to Health in Body, Mind and Intimate Relationships.* Penguin Books, 2002.

Harris, James C. "Social Neuroscience, Empathy, Brain Integration and Neurodevelopmental Disorders." *Physiology & Behavior*, vol. 79 (2003).

Harrison, Y., and J. A. Horne. "The Impact of Sleep Loss on Decision Making: A Review." *Journal of Experimental Psychology Applied*, vol. 6 (2000).

Hays, Sharon. *The Cultural Contradictions of Motherhood.* Yale University Press, 1998.

Heinrichs, Markus, et al. "Social Support and Oxytocin Interact to Suppress Cortisol and Subjective Responses to Psychosocial Stress." *Society of Biological Psychiatry* (2003).

Helgeson, Sally. *The Female Advantage.* Currency, 1995.

Hochschild, Arlie. *The Time Bind: When Work Becomes Home and Home Becomes Work.* Henry Holt & Company, 1998.

Holloway, Marguerite. "The Mutable Brain." *Scientific American* (Sept. 2003).

Holst S., et al. "Postnatal Oxytocin Treatment and Postnatal Stroking of Rats Reduce Blood Pressure in Adulthood." *Autonomic Neuroscience*, vol. 99, no. 2 (Aug. 2002).

Hondagneu-Sotelo, Pierette. "Transnational Motherhood." Essay published by Network News, a national network for an immigrant and refugee rights archive, 1997.

Hrdy, Sarah Blaffy. *Mother Nature: Maternal Instincts and How They Shape the Human Species.* Ballantine Books, 1999.

______. "Mothers and Others," *Natural History Magazine* (May 2001).

Hrdy, Sarah Blaffer, and Sue C. Carter. "Mothering and Oxytocin, or Hormonal Cocktails for Two." *Natural History* (Dec. 1995).

Jamner, Larry, et al. Paper presented at the annual scientific meetings of the Society for Behavioral Medicine, 1999.

Janowsky, J. S. "The Role Of Ovarian Hormones In Preserving Cognition In Aging." *Current Psychiatry Reports,* vol. 4, no. 6. (Dec. 2002).

Jetter, Alexis, Annelise Orleck, and Diana Taylor, eds. *The Politics of Motherhood: Activist Voices from Left to Right.* University Press of New England, 1997.

Jezova, D., et al. "Neuroendocrine Response During Stress with Relation to Gender Differences." *Acta Neurobiol. Exp.*, vol. 56, no. 3 (1996).

Johnson, Steven. *Mind Wide Open: Your Brain and the Neuroscience of Everyday Life.* Scribner, 2004.

Kane, Daniel. "How Your Brain Handles Love and Pain: Scanners Reveal Mechanisms Behind Empathy and Placebo Effect." *Science* (Feb. 19, 2004).

Kaufman, Sue. *Diary of a Mad Housewife.* Bantam Books, 2000.

Keenan, P. A., et al. "Explicit Memory in Pregnant Women." *American Journal of Obstetrics and Gynecology,* vol. 179 (1998).

Kinsley, C. H., J. Gatewood, M. Morgan, L. Flores, A. Hoffman, M. Eaten, J. Dallam, and I. McNamara. "Maternal Experience and/or Pregnancy Preserve Cognitive Performance in Aged (24-Month-Old) Female Rats." Poster presented at the annual meeting of the Society for Neuroscience, Orlando, Florida, Nov. 2002.

Kinsley, C. H., L. Madonia, G. W. Gifford, K. Tureski, G. R. Griffin, C. Lowry, J. Williams, J. Collins, and H. McLearie. "Motherhood Improves Learning and Memory," *Nature,* vol. 6758, no. 402. (Nov. 11, 1999).

Lambert, K. G. "The Life and Career of Paul Maclean: A Journey Toward Neurobiological and Social Harmony." *Physiology & Behavior,* vol. 79, no. 3 (Aug. 2003).

Lamott, Anne. *Operating Instructions: A Journal of My Son's First Year.* Ballantine, 1994.

Leckman, J. F., et al. "Early Parental Preoccupations and Behaviors and Their Possible Relationship to the Symptoms of Obsessive-Compulsive Disorder." *Acta Psychiatrica Scandinavica,* vol. 100, no. 396 (1999).

Leckman, James, and Amy Herman. "Maternal Behavior and Developmental Psychopathology." *Society of Biological Psychiatry,* vol. 51, no. 1 (2002).

Lewis, Michael, and Leonard Rosenblum. *The Effect of the Infant on Its Caregiver.* John Wiley & Sons, 1974.

Lopes, Paulo, et al. "Emotional Intelligence in the Workplace: Evidence That Emotional Intelligence Is Related to Job Performance, Interpersonal Facilitation, Affect and Attitudes At Work, and Leadership Potential." Unpublished ms., Yale University, 2004.

Lorberbaum, J. P., J. D. Newman, A. R Horwitz, J. R. Dubno, R. B. Lydiard, M. B. Hamner, D. E. Bohning, and M. S. George. "A Potential Role for Thalamocingulate Circuitry in Human Maternal Behavior." *Biol. Psychiatry*, vol. 6, no. 51.

Lorberbaum J. P., S. Kose, J. R. Dubno, A. R. Horwitz, J. D. Newman, L. Sullivan, M. B. Hamner, D. E. Bohning, G. W. Arana, and M. S. George. "Regional Brain Activity in Parents Listening to Infant Cries." Poster presented at the annual meeting of the Society for Neuroscience, San Diego, Oct. 23–27, 2004.

Lord, Janice. "Really MADD: Looking Back at 20 Years." *Driven Magazine* (Spring 2000).

Maas, James B. "Power Sleep: The Revolutionary Program That Prepares Your Mind for Peak Performance." *HarperPerennial* (1998).

MacLean, P. D. *The Triune Brain in Evolution: Role in Paleocerebral Functions.* Plenum, 1990.

______. "Women: A More Balanced Brain." *Zygon,* vol. 31, no. 3 (1996).

Maguire, E. A., et al. "Navigation-Related Structural Change in the Hippocampi of Taxi Drivers." *Proceedings of the National Academy of Sciences,* vol. 97, no. 8 (Apr. 11, 2000).

Martin, Nina. "Multitasking Makes You Sick." *Organic Style* (Nov./Dec. 2003).

Masataka N. "Perception of Motherese in Japanese Sign Language by 6-Month-Old Hearing Infants." *Developmental Psychology,* vol. 34, no. 2 (March 1998).

Mason, Mary Ann, and Marc Goulden. "Do Babies Matter? The Effect of Family Formation on the Lifelong Careers of Academic Men and Women." *Academe* (Nov.–Dec. 2002).

Mastrogiacomo, Ismaele, et al. "Postpartum Hostility and Prolactin." *International Journal of Psychiatry in Medicine,* vol. 12, no. 4 (1982).

Matloff, Judith. "Mothers At War: Babies or Battle Zones? More Journalists These Days Are Choosing Both, and Facing the Consequences." *Columbia Journalism Review* (July 2004).

Maule, Francis, et al., eds. *"The Blue Book": Women's Suffrage, History, Arguments and Results.* National Woman Suffrage Publishing Co., Inc., 1917.

McAdams, Dan, and Ed de St. Aubin. "A Theory of Generativity and its Assessment through Self-report, Behavioral Acts, and Narrative Themes in Autobiography." *Journal of Personality and Social Psychology,* vol. 62 (1992).

McKeering, Helen. "Gender and Generativity Issues in Parenting: Do Fathers Benefit More than Mothers From Involvement in Child Care Activities?" *Sex Roles: A Journal of Research* (Oct. 2000).

Merriam-Webster's Collegiate Dictionary. Tenth ed. Merriam-Webster, Inc., 1997.

Monks, D., et al. "Got Milk? Oxytocin Triggers Hippocampal Plasticity." *Nature Neuroscience* (April 2003).

Motluk, A. "Scent of a Man." *New Scientist* (Feb. 10, 2001).

Nash, J. Madeleine. "Fertile Minds: From Birth, a Baby's Brain Cells Proliferate Wildly, Making Connections That May Shape a Lifetime of Experience. The First Three Years Are Critical." *Time* (Feb. 3, 1997).

Nietzsche, Friedrich Wilhelm. *Twilight of the Idols or How to Philosophize with a Hammer.* Oxford University Press, 1998.

Nissen, H. W. "A Study of Maternal Behavior in the White Rat By Means of the Obstruction Method." *Journal of Genetic Psychology* 37 (1930).

Numan, Michael, and Thomas Insel. *The Neurobiology of Parental Behavior.* Springer, 2003.

Oatridge, Angela, et al. "Change in Brain Size During and After Pregnancy: Study in Healthy Women and Women with Preeclampsia." *American Journal of Neuroradiology,* vol. 23 (Jan. 2002).

Ochsner, Kevin N., and James J. Gross. "Thinking Makes It So: A Social Cognitive Neuroscience Approach to Emotion Regulation." Forthcoming in *The Handbook of Self-Regulation,* edited by K. Vohs and R. Baumeister. Erlbaum.

Paris, Ruth, and Ravenna Helson. "Early Mothering Experience and Personality Change." *Journal of Family Psychology,* vol. 16, no. 2 (2002).

Park, Alice. "Old Brains, New Tricks." *Time, vol.* 156, no. 6. (Aug. 7, 2000).

Pearson, Allison. *I Don't Know How She Does It: The Life of Kate Reddy, Working Mother.* Alfred Knopf, 2002.

Peri, Camille, and Kate Moses (eds). *Mothers Who Think: Tales of Real-Life Parenthood.* Villard Books, 1999.

Pilcher, Helen. "Trust Begets Hormone: Oxytocin May Help Humans Bond." *Nature Science Update* (Nov. 11, 2003).

Porter, R. H., et al. "Maternal Recognition of Neonates Through Olfactory Cues." *Physiolology & Behavior,* vol. 1, no. 1 (Jan. 1983).

Preston, Stephanie D., and Frans B. M. de Waal. "Empathy: Its Ultimate and Proximate Bases." *Behavioral and Brain Sciences,* vol. 25 (2002).

Quindlen, Anna. "Flown Away, Left Behind." *Newsweek* (Jan. 12, 2004).

Ratey, John. *A User's Guide to the Brain: Perception, Attention, and the Four Theaters of the Brain.* Vintage Books, 2001.

Reynolds, J. L. "Post-Traumatic Stress Disorder After Childbirth: The Phenomenon of Traumatic Birth. *Canadian Medical Association Journal,* vol. 156 (1997).

Rilling, J., et al. "A Neural Basis for Social Cooperation." *Neuron,* vol. 35, no. 2 (July 2002)

Roberts, R. L., et al. "Prolactin Levels Are Elevated After Infant Carrying In Parentally Inexperienced Common Marmosets." *Physiolology & Behavior,* vol. 72, no. 5 (Apr. 2001).

Rosenthal, Robert, et al. *Sensitivity to Nonverbal Communication: The PONS Test.* The Johns Hopkins University Press, 1979.

Rosenwein, Rifka. "More Moms Stay Home." *Forecast* (Nov. 19, 2001).

Rowe-Finkbeiner, Kristin. "Juggling Career and Home: Albright, O'Connor, and You." *Mothering Magazine,* no. 117 (March/April 2003).

Ruddick, Sara. *Maternal Thinking: Toward a Politics of Peace.* London: The Women's Press Ltd., 1990.

______."What Do Grandmothers Know and Want?" Essay in *What Do Mothers Want?* The Analytic Press, forthcoming.

______. "Making Connections Between Parenting and Peace." *Journal of the Association for Research on Mothering,* vol. 3, no. 2 (Fall/Winter 2001).

Sagan, Carl. *The Dragons of Eden: Speculations on the Evolution of Human Intelligence.* Ballantine Books, 1977.

Sapolsky, Robert. "Stress and Cognition." Paper prepared for *The Cognitive Neurosciences.* MIT Press, 2004.

Scheffler, H. W., and F. G. Lounsbury. *A Study in Structural Semantics: The Siriono Kinship System.* Prentice-Hall, 1971.

Schroeder, Pat. *24 Years of Housework . . . and the Place Is Still a Mess.* Andrews McMeel, 1998.

Seifritz, E., et al. "Differential Sex-Independent Amygdala Response to Infant Crying and Laughing in Parents Versus Nonparents." *Biological Psychiatry,* vol. 54, no. 12 (Dec. 15, 2003).

Senge, Peter M., and Goran Carstedt. "Innovating Our Way to the Next Industrial Revolution." *MIT Sloan Management Review,* Winter 2001.

Sherwin, B. B. "Estrogen and Cognitive Aging in Women." *Trends in Pharmacological Science,* vol. 23, no. 11 (Nov. 2002).

Shingo, T., et al. "Pregnancy-Stimulated Neurogenesis in the Adult Female Forebrain Mediated by Prolactin." *Science,* vol. 299, no. 5603 (Jan. 3, 2003).

Silk, J. B., et al. "Social Bonds of Female Baboons Enhance Infant Survival." *Science,* vol. 302, (Nov. 14, 2003).

Singer, Mark. "Mom Overboard! What Do Power Women Who Decide to Quit the Fast Lane Do with Themselves All Day?" *New Yorker* (Feb. 26, 1996).

Sinha, Gunjan, "You Dirty Vole." *Popular Science* (Nov. 2003).

Skocpol, Theda. *Protecting Soldiers and Mothers: The Political Origins of Social Policy in the United States.* Harvard University Press, 1995.

Small, Meredith. "Family Matters: Nature and Nurture in Dominica." *Discover* (Aug. 2000).

Smith, Janna Malamud. *A Potent Spell: Mother Love and the Power of Fear.*" Houghton Mifflin, 2003.

Stadtman Tucker, Judith. "An Interview with Enola Aird." *Mothers Movement Online* (June 2003).

______."The Least Worst Choice: Why Mothers 'Opt' Out of the Workforce." *Mothers Movement Online* (Dec. 2003).

Stanton, Elizabeth Cady. *Elizabeth Cady Stanton As Revealed in Her Letters, Diary, and Reminiscences.* Vol. 1. Harper & Row, 1922.

Stearns, Peter N. *Anxious Parents: A History of Modern Childrearing in America.*" New York University Press, 2003.

Stern, Daniel N. *The Birth of a Mother: How the Motherhood Experience Changes You Forever.* Basic Books, 1998.

Tanapat, Patima, et al. "Estrogen Stimulates a Transient Increase in the Number of New Neurons in the Dentate Gyrus of the Adult Female Rat." *Journal of Neuroscience,* vol. 19, no. 14 (July 15, 1999).

Tang, Y., et al. "Estrogen Replacement Increases Spinophilin-Immunoreactive Spine Number in the Prefrontal Cortex of Female Rhesus Monkeys." *Cerebral Cortex,* vol. 14, no. 2 (Feb. 2004).

Taylor, Shelley E., et al. "Behavioral Responses to Stress in Females: Tend-and-Befriend, Not Fight-or-Flight." *Psychological Review* 107, no. 3 (2000).

______. *The Tending Instinct: How Nurturing Is Essential to Who We Are and How We Live.* Times Books, Henry Holt and Company, 2002.

Theodosis, D. T., and D. A. Poulain. "Maternity Leads to Morphological Synaptic Plasticity in the Oxytocin System." *Progress in Brain Research,* vol. 133 (2001).

Thoreau, Henry. *The Essays of Henry D. Thoreau.* Edited by Lewis Hyde. North Point Press, 2002.

Tomizawa, Kazuhito, et al. "Oxytocin Improves Long-Lasting Spatial Memory During Motherhood Through MAP Kinase Cascade." *Nature Neuroscience,* vol. 6, no 4 (April 2003).

Torner, L., et al. "Anxiolytic and Anti-Stress Effects of Brain Prolactin: Improved Efficacy of Antisense Targeting of the Prolactin Receptor by Molecular Modeling." *Journal of Neuroscience* 21, no. 9 (May 1, 2001).

Tronick, E. Z. "Infant Moods and the Chronicity of Depressive Symptoms: The Co-Creation of Unique Ways of Being Together for Good or Ill." Forthcoming in *Zeitschrift fur Psychosomatische Medizin und Psychotherapie,* 2003.

Tumulty, Karen, and Viveca Novak. "Goodbye, Soccer Mom, Hello, Security Mom." *Time* (May 27, 2003).

Uvnas-Moberg, Kerstin. *The Oxytocin Factor: Tapping the Hormone of Calm, Love and Healing.* Da Capo Press, 2003.

______. "Oxytocin May Still Mediate the Benefits of Positive Social Interaction and Emotions." *Psychoneuroendocrinology,* vol. 23 (1998).

Vaillant, George. *Aging Well: Surprising Guideposts to a Happier Life from the Landmark Harvard Study of Adult Development.* Little, Brown, 2003.

Vanston, Claire, and Neil Watson. "Persistent Effect of Fetal Sex on Maternal Cognitive Function: A Longitudinal Study in Pregnant Women." Unpublished paper, Simon Fraser University, 2004.

Waldman, Ayelet. *Nursery Crimes.* Berkley Publishing Group, 2001.

Wallis, Claudia. "The Case for Staying Home." *Time* (March 22, 2004).

______. "What Makes Teens Tick." *Time* (May 10, 2004).

Wang, Z.X., Y. Liu, L. J. Young, and T. R. Insel. "Hypothalamic Vasopressin Gene Expression Increases in Both Males and Females Postpartum in a Biparental Rodent." *Journal of Neuroendocrinology,* vol. 12, no. 2 (Feb. 2000).

Ward, Geoffrey, and Ken Burns. *Not for Ourselves Alone.* Alfred A. Knopf, 1999.

Wartella, J., et al. "Single or Multiple Reproductive Experiences Attenuate Neurobehavioral Stress and Fear Responses in the Female Rat." *Physiology & Behavior,* vol. 79 (2003).

Weldon, Fay. *Puffball.* Pocket Books, 1980.

"Why Do Humans and Apes Cradle Babies on Their Left Side?" *New Scientist* 127, no. 1726 (July 21, 1990).

Woolley, C. S., and B. McEwen. "Estradiol Mediates Fluctuation in Hippocampal Synapse Density During the Estrous Cycle in the Adult Rat." *Journal of Neuroscience,* vol. 12, no. 10 (Oct. 1992).

Xerri, C., et al. "Alterations of the Cortical Representation of the Rat Ventrum Induced by Nursing Behavior." *Journal of Neuroscience,* vol. 14, no. 3, pt. 2 (March 1994).

INDEX